4995

Theory of Difference Equations

Numerical Methods and Applications

This is a volume in
MATHEMATICS IN SCIENCE AND ENGINEERING
Edited by William F. Ames, Georgia Institute of Technology
A list of recent titles in this series appears at the end of this volume.

Theory of Difference Equations
Numerical Methods and Applications

V. Lakshmikantham
Department of Mathematics
University of Texas at Arlington
Arlington, Texas

D. Trigiante
Dipartimento di Matematica
Università di Bari
Bari, Italy

ACADEMIC PRESS, INC.
Harcourt Brace Jovanovich, Publishers
Boston San Diego New York
Berkeley London Sydney
Tokyo Toronto

ACADEMIC PRESS, INC.
1250 Sixth Avenue, San Diego, CA 92101

United Kingdom Edition published by
ACADEMIC PRESS, INC. (LONDON) LTD.
24–28 Oval Road, London NW1 7DX

Library of Congress Cataloging-in-Publication Data

Lakshmikantham, V.
Theory of difference equations.

(Mathematics in science and engineering)
Includes index.
Bibliography: p.
1. Difference equations. 2. Numerical analysis.
I. Trigiante, D. II. Title.
QA431.L28 1987 515′.625 87-16783
ISBN 0-12-434100-4

88 89 90 91 9 8 7 6 5 4 3 2 1
Printed in the United States of America

Contents

CHAPTER 3 Linear Systems of Difference Equations

CHAPTER 4 Stability Theory

CHAPTER 5 Applications to Numerical Analysis

Preface

Difference equations appear as natural descriptions of observed evolution phenomena because most measurements of time evolving variables are discrete and as such these equations are in their own right important mathematical models. More importantly, difference equations also appear in the study of discretization methods for differential equations. Several results in the theory of difference equations have been obtained as more or less natural discrete analogues of corresponding results of differential equations. This is especially true in the case of Lyapunov theory of stability. Nonetheless, the theory of difference equations is a lot richer than the corresponding theory of differential equations. For example, a simple difference equation resulting from a first order differential equation may have a phenomena often called appearance of "ghost" solutions or existence of chaotic orbits that can only happen for higher order differential equations. Consequently, the theory of difference equations is interesting in itself and it is easy to see that it will assume greater importance in the near future. Furthermore, the application of the theory of difference equations is rapidly increasing to various fields such as numerical analysis, control theory, finite mathematics and computer science. Thus, there is every reason for studying the theory of difference equations as a well deserved discipline.

The present book offers a systematic treatment of the theory of difference equations and its applications to numerical analysis. It does not treat in detail the classical applications of difference calculus to numerical analysis such as interpolation theory, numerical quadrature and differentiation, which can be found in many classical books. Instead, we devote our attention to iterative processes and numerical methods for differential equations. The investigation of these subjects from the point of view of difference equations allows us to systematize and clarify the ideas involved and, as a result, pave the way for further developments of this fruitful union.

The book is divided into seven chapters. The first chapter introduces difference calculus, deals with preliminary results on difference equations and develops the theory of difference inequalities. In the second chapter, we present the essential techniques employed in the treatment of linear difference equations with special reference to equations with constant coefficients. Chapter 3 deals with the basic theory of systems of linear difference equations. Chapter 4 is devoted to the theory of stability of Lyapunov including converse theorems and total and practical stability. Chapters 5 and 6 discuss the application of the theory of difference equations to numerical analysis. In Chapter 7, we present applications of difference equations to many fields such as economics, chemistry, population dynamics and queueing theory. The necessary linear algebra used in the book is given in Appendices. Finally, several carefully selected problems at the end of each chapter complement the material of the book.

Some of the important features of the book include the following:

(i) development of the theory of difference inequalities and the various comparison results;
(ii) a unified treatment of stability theory through Lyapunov functions and comparison method;
(iii) stressing the important role of the theory of difference equations in numerical analysis;
(iv) demonstrating the versatility of difference equations by various models of the real world;
(v) timely recognition of the importance of the theory of difference equations and of presenting a unified treatment.

The book can be used as a textbook at the graduate level and as a reference book. Readers interested in discrete dynamical systems will also find the material, which is not available in other books on difference equations, useful.

We wish to express our immense thanks to Professors Ilio Galligani, Vincenzo Casulli, Giuseppe Piazza and S. Leela for their helpful comments and suggestions. We are happy to convey our thanks to Ms. Sandra Weber for her excellent typing of the manuscript at several stages. We express our appreciation and thanks to Mr. William Sribney, Editor, Academic Press, for his helpful cooperation.

CHAPTER 1 Preliminaries

1.0. Introduction

This chapter is essentially introductory in nature. Its main aim is to introduce certain well-known basic concepts in difference calculus and to present some important results that are not as well-known. Sections 1.1 to 1.4 contain needed difference calculus and some notions related to it, most of which is found in standard books on difference equations. Section 1.5 deals with preliminary results on difference equations such as existence and uniqueness and also discusses solving some simple difference equations. In Section 1.6, we develop theory of difference inequalities and prove a variety of comparison theorems that play a prominent role in the development of the book. Several problems are given in Section 1.7 that, together with the material of Sections 1.1 to 1.4, cover the necessary theory of difference calculus.

1.1. Operators Δ and E

We shall consider functions defined on a discrete set of points $J^+_{x_0}$ or $J^+_{x'_0,h}$ given by

$$J^+_{x_0} = \{x_0, x_0 + 1, \ldots, x_0 + k, \ldots\}.$$

$$J^+_{x'_0,h} = \{x'_0, x'_0 + h, \ldots, x'_0 + kh, \ldots\},$$

where $x_0' = hx_0 \in \boldsymbol{R}$ and $x_0 \in \boldsymbol{R}$ (sometimes $x_0 \in \boldsymbol{C}$). Here $\boldsymbol{R}$ and $\boldsymbol{C}$ denote as usual, the real and complex numbers respectively. The set $\boldsymbol{N}^+$ of natural numbers that is isomorphic to both $J_{x_0}^+$, $J_{x_0',h}^+$ is also used as the definition set when it is desirable to exhibit the dependence of the function on $n \in \boldsymbol{N}^+$. If $y : \boldsymbol{N}^+ \to \boldsymbol{C}$ is the discrete function, $y(n)$ is the sequence denoted by y_n. If the discrete function represents the values of a continuous function at discrete points, then the set $J_{x_0}^+$ or $J_{x_0',h}^+$ is preferred since in this case, the dependence on x_0 is shown explicitly. The advantage in using $J_{x_0',h}^+$ as the definition set is that, for a discrete function which is the approximation of a function defined on $\boldsymbol{R}$, the dependence of the approximation error on the step h is indicated. The correspondence between functions defined on $\boldsymbol{N}^+$ and $J_{x_0}^+$ is given by $y_n = y(n) = y(x_0 + n)$, where y is the discrete function with values in $\boldsymbol{R}$ or $\boldsymbol{C}$.

In some cases, the following sets are also used:

$$J_{x_0,h}^{\pm} = \{x_0, x_0 \pm h, \ldots, x_0 \pm kh, \ldots\},$$

$$J_{x_0}^{\pm} = \{x_0, x_0 \pm 1, \ldots, x_0 \pm k, \ldots\},$$

$$\boldsymbol{N} = \{0, \pm 1, \pm 2, \ldots, \pm k, \ldots\},$$

$$\boldsymbol{N}_{n_0}^{\pm} = \{n_0, n_0 \pm 1, \ldots, n_0 \pm k, \ldots\},$$

with $k \in \boldsymbol{N}^+$ and $n_0 \in \boldsymbol{N}$.

In this chapter, we shall often use J_x^+ as the definition set whenever we need the dependence on $x \in \boldsymbol{R}$ (or $x \in \boldsymbol{C}$) in order to consider derivatives (or differences) with respect to x. There will be no difficulty in translating the results in terms of other notations. As a rule, we shall only use the sequence notation in the problems at the end of chapters.

Definition 1.1.1. Let $y : J_x^+ \to \boldsymbol{C}$. Then $\Delta : \boldsymbol{C} \to \boldsymbol{C}$ is the difference operator defined by

$$\Delta y(x) = y(x+1) - y(x) \tag{1.1.1}$$

and $E : \boldsymbol{C} \to \boldsymbol{C}$ is the shift operator defined by

$$Ey(x) = y(x+1). \tag{1.1.2}$$

It is easy to verify that the two operators Δ and E are linear and that they commute. That is, for any two functions y, z and J_x^+ and any two scalars α, β, we have

$$\Delta(\alpha y(x) + \beta z(x)) = \alpha \Delta y(x) + \beta \Delta z(x),$$

$$E(\alpha y(x) + \beta z(x)) = \alpha E y(x) + \beta E z(x),$$

and $\Delta E y(x) = E \Delta y(x)$.

The second difference on $y(x)$ can be defined as

$$\Delta^2 y(x) = \Delta(\Delta y(x)) = y(x+2) - 2y(x+1) + y(x).$$

In general, for every $k \in N^+$,

$$\Delta^k y(x) = \Delta(\Delta^{k-1} y(x) \quad \text{and} \quad E^k y(x) = y(x+k),$$

with $\Delta^0 y(x) = E^0 y(x) = Iy(x)$, I being the identity operator such that $Iy(x) = y(x)$. In the case when the definition set is N^+, one has $\Delta y_n = y_{n+1} - y_n$ and $Ey_n = y(n+1)$ for (1.1.1), (1.1.2) respectively. It is easy to see that the formal relation between Δ and E is $\Delta = E - I$ and thus, powers of Δ can be expressed in terms of powers of E and vice versa. In fact,

$$\Delta^k = (E - I)^k = \sum_{i=0}^{k} (-1)^{k-i} \binom{k}{i} E^i, \tag{1.1.3}$$

and

$$E^k = (\Delta + I)^k = \sum_{i=0}^{k} \binom{k}{i} \Delta^i, \tag{1.1.4}$$

where $\binom{k}{i}$ are the binomial coefficients.

Definition 1.1.2. Let $y : J_x^+ \to C$. The function y is said to be periodic of period k if $y(x+k) = y(x)$.

The constant functions are particular periodic functions and $y(x) = e^{i2\pi x}$ is a periodic function of period 1. It is easy to see that $\Delta y(x) = 0$ for any periodic function of period 1.

1.2. Negative Powers of Δ

Consider the equation

$$\Delta y(x) = g(x), \tag{1.2.1}$$

where $y, g : J_x^+ \to C$. The solution of this equation is a function y such that (1.2.1) is satisfied. We shall denote this solution by $y(x) = \Delta^{-1} g(x)$. It is not unique because $y(x)$ can also be expressed as $\Delta^{-1} g(x) + \omega(x)$, where $\omega(x)$ is an arbitrary function of period 1. The operator Δ^{-1} is called the antidifference and it is linear. Moreover, the operators Δ and Δ^{-1} do not commute since

$$\Delta\Delta^{-1} = I \quad \text{and} \quad \Delta^{-1}\Delta = I + \omega(x).$$

If $f, g : J_x^+ \to C$ are two functions such that $\Delta f(x) = \Delta g(x)$, then it is clear that $f(x) = g(x) + \omega(x)$. In particular, if $f(x)$ and $g(x)$ are polynomials, $\Delta f = \Delta g$ implies $f(x) = g(x) + c$, where c is a constant.

We shall now state a fundamental result that enables us to compute the finite sum $\sum_{i=0}^{n} f(x+i)$ in terms of the antidifference $\Delta^{-1} f(x)$.

Theorem 1.2.1. *Let $F(x) = \Delta^{-1} f(x)$. Then*

$$\sum_{i=0}^{n} f(x+i) = F(x+n+1) - F(x) \equiv F(x+i)\Big|_{i=0}^{i=n+1}. \tag{1.2.2}$$

Proof. Since by hypothesis we have $f(x) = \Delta F(x)$, it is easy to see that

$$\begin{aligned}\sum_{i=0}^{n} f(x+i) &= \sum_{i=0}^{n} \Delta F(x+i)\\ &= \sum_{i=0}^{n} [F(x+i+1) - F(x+i)]\\ &= F(x+n+1) - F(x). \quad \blacksquare\end{aligned}$$

Note that (1.2.2) can also be written as

$$\sum_{i=0}^{n} f(x+i) = \Delta^{-1} f(x+i)\Big|_{i=0}^{i=n+1}. \tag{1.2.3}$$

If we leave the sum to remain indefinite, we can express the relation (1.2.3) in the form

$$\sum f(x) = \Delta^{-1} f(x) + \omega(x)$$

in analogy with the notation for indefinite integrals.

In the case when the definition set is N^+, the foregoing formulas reduce to

$$\sum_{i=0}^{n} y_i = \Delta^{-1} y_i\Big|_{i=0}^{i=n+1} \quad \text{and} \quad \sum y_i = \Delta^{-1} y_i + \omega$$

respectively.

1.3. Factorial Powers and Discrete Taylor Formulas

Analogous to the role of functions x^n in differential and integral calculus, one has the factorial powers of x, denoted by $x^{(n)}$.

Definition 1.3.1. Let $x \in \boldsymbol{R}$. The nth factorial power of x is defined by

$$x^{(n)} = x(x-1)\dots(x-n+1).$$

It is easy to verify from the above definition that

$$\Delta x^{(n)} = nx^{(n-1)} \quad \text{and} \quad \Delta^{-1}x^{(n-1)} = \frac{1}{n}x^{(n)} + \omega(x). \tag{1.3.1}$$

We also have $x^{(m+n)} = x^{(m)}(x-m)^{(n)} = x^{(n)}(x-n)^{(m)}$. For $m = 0$, this yields $x^{(0+n)} = x^{(0)}x^{(n)}$, which shows that $x^{(0)} = 1$. Moreover, for $m = -n$, we get $1 = x^{(0)} = x^{(-n)}(x+n)^{(n)}$, which allows one to define the negative factorial power $x^{(-n)}$ by

$$x^{(-n)} = \frac{1}{(x+1)(x+2)\dots(x+n)}.$$

The relations (1.3.1) suggest that it will be convenient to express other functions in terms of factorial powers whenever possible. For example, in the case of polynomials we have the following result.

Theorem 1.3.1. *The powers x^n and the factorial powers are related by*

$$x^n = \sum_{i=1}^{n} S_i^n x^{(i)}, \tag{1.3.2}$$

where S_i^n are the Sterling numbers (of second kind) that satisfy the relation

$$S_i^{n+1} = S_{i-1}^n + iS_i^n$$

with $S_n^n = S_1^n = 1$.

Proof. Clearly (1.3.2) holds for $n = 1$. Suppose it is true for some n, multiplying both sides of (1.3.2) by x, we get

$$\begin{aligned}
x^{n+1} &= \sum_{i=1}^{n} S_i^n x \cdot x^{(i)} = \sum_{i=1}^{n} S_i^n x(x-1)\dots(x-i+1)(x-i+i) \\
&= \sum_{i=1}^{n} iS_i^n x^{(i)} + \sum_{i=1}^{n} S_i^n x^{(i+1)} \\
&= \sum_{i=2}^{n} [iS_i^n + S_{i-1}^n]x^{(i)} + S_1^{n+1}x^{(1)} + S_{n+1}^{n+1}x^{(n+1)} \\
&= \sum_{i=1}^{n+1} S_i^{n+1}x^{(i)},
\end{aligned}$$

showing that (1.3.2) holds for $n+1$. Hence the proof is complete by induction. ■

Sterling numbers S_i^n for $i, n = 1, 2, \dots, 6$ are given in Table 1.

TABLE 1. Sterling numbers of Second Kind

n^i	1	2	3	4	5	6
1	1					
2	1	1				
3	1	3	1			
4	1	7	6	1		
5	1	15	25	10	1	
6	1	31	90	65	15	1

Using the relation (1.3.2), it is immediate to derive the differences and the antidifferences of a polynomial.

Theorem 1.3.2. *The first difference of a polynomial of degree k is a polynomial of degree k − 1 and in general, the sth difference is a polynomial of degree k − s.*

Proof. It is not restrictive to consider x^k. From (1.3.2) we have

$$\Delta^s x^k = \sum_{i=1}^{k} S_i^k \Delta^s x^{(i)} = \sum_{i=1}^{k} i^{(s)} S_i^k x^{(i-s)},$$

which is of degree $k - s$.

It is easy to check that for $x \in \mathbf{N}_0^+$, $\dfrac{x^{(j)}}{j!} = \dbinom{x}{j}$ and that $\Delta\dbinom{x}{j} = \dbinom{x}{j-1}$, $j \geq 1$. We shall now establish some discrete analogues of the Taylor formula. Using the formula (1.1.4), we obtain the first results. ■

Theorem 1.3.3. *Let $n \in \mathbf{N}_0^+$ and u_n be defined on $\mathbf{N}_0^+$. Then*

$$u_n = E^n u_0 = \sum_{i=0}^{n} \binom{n}{i} \Delta^i u_0. \tag{1.3.3}$$

Theorem 1.3.4. *Let k, $n \in \mathbf{N}_0^+$, $k \leq n$ and u_n be defined on $\mathbf{N}_0^+$. Then*

$$u_n = \sum_{i=0}^{k-1} \binom{n}{i} \Delta^i u_0 + \sum_{s=0}^{n-k} \binom{n-s-1}{k-1} \Delta^k u_s. \tag{1.3.4}$$

Proof. From (1.3.3) it follows that

$$\begin{aligned} u_n &= \sum_{i=0}^{k-1} \binom{n}{i} \Delta^i u_0 + \sum_{j=0}^{n-k} \binom{n}{k+j} \Delta^{k+j} u_0 \\ &= \sum_{i=0}^{k-1} \binom{n}{i} \Delta^i u_0 + \sum_{j=0}^{n-k} \binom{n}{k+j} \Delta^k \sum_{s=0}^{j} (-1)^{j-s} \binom{j}{s} E^s u_0 \\ &= \sum_{i=0}^{k-1} \binom{n}{i} \Delta^i u_0 + \sum_{s=0}^{n-k} \left[\sum_{j=s}^{n-k} (-1)^{j-s} \binom{n}{k+j} \binom{j}{s} \right] \Delta^k E^s u_0. \end{aligned}$$

■

By using the identity (see problem 1.11)

$$\sum_{j=s}^{n-k} (-1)^{j-s}\binom{n}{k+j}\binom{j}{s} = \binom{n-s-1}{k-1},$$

one obtains (1.3.4), which is called discrete Taylor formula. A generalization of Theorem 1.3.4 is as follows.

Theorem 1.3.5. *Let $j, k, n \in \mathbf{N}_0^+$, $j \leq k-1$, $k \leq n$ and u_n be defined on $\mathbf{N}_0^+$. Then*

$$\Delta^j u_n = \sum_{i=j}^{k-1}\binom{n}{i-j}\Delta^i u_0 + \sum_{s=0}^{n-k+j}\binom{n-s-1}{k-j-1}\Delta^k u_s. \tag{1.3.5}$$

Proof. For $j = 0$, (1.3.5) reduces to (1.3.4). Suppose that (1.3.5) holds for some j. Then, one has

$$\begin{aligned}
\Delta^{j+1} u_n &= \sum_{i=j+1}^{k-1}\binom{n}{i-j-1}\Delta^i u_0 + \sum_{s=0}^{n-k+j}\binom{n-s-1}{k-j-2}\Delta^k u_s \\
&\quad + \binom{k-j-1}{k-j-1}\Delta^k u_{n+1-k+j} \\
&= \sum_{i=j+1}^{k-1}\binom{n}{i-j-1}\Delta^i u_0 + \sum_{s=0}^{n-k+j}\binom{n-s-1}{k-j-2}\Delta^k u_s \\
&\quad + \binom{k-j-2}{k-j-2}\Delta^k u_{n+1-k+j} \\
&= \sum_{i=j+1}^{k-1}\binom{n}{i-j-1}\Delta^i u_0 + \sum_{s=0}^{n-k+j+1}\binom{n-s-1}{k-j-2}\Delta^k u_s,
\end{aligned}$$

which is (1.3.5) for $j+1$. ■

1.4. Bernoulli Numbers and Polynomials

From (1.3.2) one has that for every $n \in \mathbf{N}^+$, $x \in \mathbf{R}$

$$\Delta^{-1} x^n = \sum_{i=1}^{n} S_i^n \frac{x^{(i+1)}}{i+1} + \omega_n(x), \tag{1.4.1}$$

where $\omega_n(x)$ are periodic functions. If we require $\Delta^{-1}x^n$ to be a polynomial, then we must choose $\omega_n(x)$ as constants with respect to x. Let $\dfrac{1}{n+1}C_{n+1}$ be such constants, and let us write

$$B_{n+1}(x) = (n+1)\Delta^{-1}x^n = (n+1)\sum_{i=1}^{n} S_i^n \frac{x^{(i+1)}}{i+1} + C_{n+1} \tag{1.4.2}$$

with $B_0(x) = 1$. The polynomials $B_n(x)$ satisfy the relation

$$\Delta B_n(x) = nx^{n-1}. \tag{1.4.3}$$

They are not uniquely defined because the constants C_n are arbitrary. Usually it is convenient to avoid the Sterling numbers in the determination of $B_n(x)$. This can be accomplished as follows.

Theorem 1.4.1. *Let $n \in N^+$, $B_0(x) = 1$ and $B_n(x)$ be polynomials satisfying* (1.4.3). *Then the two functions*

$$F_n(x) = \sum_{i=0}^{n-1} \binom{n}{i} B_i(x), \tag{1.4.4}$$

and

$$G_n(x) = nx^{n-1}, \tag{1.4.5}$$

differ by a constant.

Proof. From (1.4.3) one has

$$\Delta F_n(x) = \sum_{i=0}^{n-1} \binom{n}{i} ix^{i-1}$$

and

$$\Delta G_n(x) = n[(x+1)^{n-1} - x^{n-1}] = \sum_{i=0}^{n-1} \binom{n}{i} ix^{i-1}.$$

Hence, it follows that $\Delta F_n(x) = \Delta G_n(x)$ and since $F_n(x)$, $G_n(x)$ are polynomials, we have

$$F_n(x) = G_n(x) + d_n, \tag{1.4.6}$$

where d_n are constants. ■

When the constants d_n have been fixed, (1.4.4) and (1.4.6) allow us to construct the polynomials $B_n(x)$. The constants d_n are fixed by imposing one more condition to be satisfied by $B_n(x)$. The most commonly used condition is

$$\frac{dB_n(x)}{dx} = nB_{n-1}(x), \tag{1.4.7}$$

or

$$\int_0^1 B_n(x)\, dx = 0, \qquad \text{for } n = 1, 2, \ldots. \tag{1.4.8}$$

One, in fact, has the following result.

Theorem 1.4.2. *If for every $n \in \mathbf{N}^+$, the polynomials $B_n(x)$ satisfy* (1.4.3) *with $B_0(x) = 1$, and either* (1.4.7) *or* (1.4.8) *is satisfied, then $d_n = 0$, and*

$$\sum_{i=0}^{n-1} \binom{n}{i} B_i(x) = nx^{n-1}. \tag{1.4.9}$$

Proof. Let us start with (1.4.7). Differentiating (1.4.6) and using (1.4.7), we have $nF_{n-1}(x) = F'_n(x) = G'_n(x) = nG_{n-1}(x)$. This implies that $0 = n(F_{n-1}(x) - G_{n-1}(x)) = nd_{n-1}$ from which it follows $d_{n-1} = 0$.

Let us now suppose that (1.4.8) holds. From (1.4.4) we obtain $\int_0^1 F_n(x)\,dx = \int_0^1 B_0(x)\,dx = 1$ and $\int_0^1 G_n(x)\,dx = 1$. Because of (1.4.6) we now get $1 = 1 + d_n$, which implies $d_n = 0$. ∎

As we have already observed, (1.4.9) and $B_0(x) = 1$ define the polynomials $B_n(x)$ uniquely, and these are called Bernoulli polynomials. The first five of these polynomials are as follows:

$$B_0(x) = 1,\ B_1(x) = x - \tfrac{1}{2},$$
$$B_2(x) = x^2 - x + \tfrac{1}{6},$$
$$B_3(x) = x^3 - \tfrac{3}{2}x^2 + \tfrac{1}{2}x,$$

and

$$B_4(x) = x^4 - 2x^3 + x^2 - \tfrac{1}{30}.$$

The values of $B_n(0)$ are called Bernoulli numbers and are denoted by B_n. As an easy consequence of (1.4.9), we see that the Bernoulli numbers satisfy the relation

$$\sum_{i=0}^{n-1} \binom{n}{i} B_i = 0, \qquad n = 2, 3, \ldots,$$

which can be considered as the expansion of $(1 + B)^n - B^n$, where the powers B^i are replaced by B_i. This property is often used to define Bernoulli numbers. It can be shown that the Bernoulli numbers of odd index, except for B_1, are zero. The values of the first ten numbers are

$$B_0 = 1,\ B_1 = -\tfrac{1}{2},\ B_2 = \tfrac{1}{6},\ B_4 = -\tfrac{1}{30},\ B_6 = \tfrac{1}{42},\ B_8 = -\tfrac{1}{30},\ B_{10} = \tfrac{5}{66}.$$

From (1.4.3), applying Δ^{-1} to both sides, we get

$$\Delta^{-1}x^{n-1} = \frac{B_n(x)}{n}. \tag{1.4.10}$$

A simple application of (1.4.10) is the following. Suppose that x takes integer values. Then, from (1.2.3) and (1.4.9), we see that

$$\sum_{x=0}^{m} x^{n-1} = \left.\frac{B_n(x)}{n}\right|_{x=0}^{x=m+1} = \frac{1}{n}[B_n(m+1) - B_n],$$

TABLE 2. Differences and Antidifferences

$f(x)$	$\Delta f(x)$	$\Delta^{-1} f(x)$
c	0	cx
c^x	$(c-1)c^x$	$\dfrac{c^x}{c-1},\ c \neq 1$
xc^x	$(c-1)xc^x + c^{x+1}$	$\dfrac{c^x}{c-1}\left(x - \dfrac{c}{c-1}\right),\ c \neq 1$
$(x+b)^{(n)}$	$n(x+b)^{(n-1)}$	$\dfrac{(x+b)^{(n+1)}}{n+1},\ n \neq -1$
$\binom{x}{n}$	$\binom{x}{n-1}$	$\binom{x}{n+1}$
$\cos(ax+b)$	$-2\sin\dfrac{a}{2}\sin\left(ax+b+\dfrac{a}{2}\right)$	$\dfrac{\sin\left(ax+b-\dfrac{a}{2}\right)}{2\sin\dfrac{a}{2}}$
$\sin(ax+b)$	$2\sin\dfrac{a}{2}\cos\left(ax+b+\dfrac{a}{2}\right)$	$\dfrac{-\cos\left(ax+b-\dfrac{a}{2}\right)}{2\sin\dfrac{a}{2}}$
x^n	$\sum_{i=0}^{n-1}\binom{n}{i}x^i$	$\dfrac{B_{n+1}(x)}{n+1},\ n \neq -1$
$\log(x+c)$	$\log\left(1+\dfrac{1}{x+c}\right)$	$\log \Gamma(x+c)$

from which we get the sum of the $(n-1)^{\text{th}}$ powers of integer numbers. When $n = 3$, for example, we have

$$\sum_{x=0}^{m} x^2 = \tfrac{1}{3}(m+1)^3 - \tfrac{3}{2}(m+1)^2 + \tfrac{1}{2}(m+1) = \tfrac{1}{6}m(m+1)(2m+1).$$

In Table 2 we list differences and antidifferences of the most common functions, omitting the periodic function $\omega(x)$.

1.5. Difference Equations

We have seen that the knowledge of $\Delta^{-1}g(x)$ allows us to solve the equation (1.2.1), which is a very simple difference equation. In general, a difference equation of order k will be a functional relation of the form

$$F(x, y(x), \Delta y(x), \ldots, \Delta^k y(x), g(x)) = 0,$$

where $y, g : J_{x_0}^{+} \to \boldsymbol{C}$.

More often, instead of the operator Δ, one uses the operator E. The difference equation is then written in the form

$$G(x, y(x), Ey(x), \ldots, E^k y(x), g(x)) = 0.$$

If the function F (or G) is linear with respect to $y(x), \Delta y(x), \ldots, \Delta^k y(x)$ (or respectively $y(x), Ey(x), \ldots, E^k y(x)$), then the difference equation is said to be linear. The theory of linear difference equations will be presented in Chapter 2.

Except for some specific cases, we shall consider, in the following, difference equations that can be written in the normal form

$$E^k y(x) = \Phi(x, y(x), Ey(x), \ldots, E^{k-1} y(x), g(x)), \tag{1.5.1}$$

where Φ is a function uniquely defined in its arguments in some subset $D \subset J^+_{x_0} \times C \times \ldots \times C$. With the equation (1.5.1), we associate k initial conditions

$$y(x) = c_1, \quad y(x+1) = c_2, \quad \ldots, \quad y(x+k-1) = c_k. \tag{1.5.2}$$

An existence and uniqueness result for the problem (1.5.1), (1.5.2) is the following.

Theorem 1.5.1. *The difference equation* (1.5.1) *with the initial conditions* (1.5.2) *has a unique solution* $y(x+n)$, *if the arguments of* Φ *are in* D.

Proof. From (1.5.1) and (1.5.2), we get $y(x+k)$. By changing x into $x+1$ in the value $y(x+k)$ and using the last $(k-1)$ values from (1.5.2), we obtain $y(x+1+k)$ in view of (1.5.1). Repetition of this procedure yields the unique solution $y(x+n)$. ■

The possibility of obtaining the values of solutions of difference equations recursively is very important and does not have a counterpart in other kinds of equations. Having at our disposal machines that can do a large number of calculations in a second, we can get, in a short time, a great number of values of the solution of difference equations. For this reason, one reduces continuous problems to approximate discrete problems. This way of obtaining solutions, however, although very efficient for some purposes, is insufficient for others. For example, it does not give information on asymptotic behavior of solutions, unless one is willing to accept costs that are exceedingly high. Hence, it is of great importance to have solutions in a closed analytical form, or, at least, to deduce information on the qualitative behavior of the solutions in some other way.

Next, we shall discuss solutions of some simple difference equations. We have already seen that the solution of $\Delta y(x) = g(x)$ is $\Delta^{-1}g(x)$. It is sometimes possible to reduce certain other equations to this form. This is the case, for example, with the equation

$$z(x+1) - p(x)z(x) = q(x), \qquad z(x_0) = z_0. \tag{1.4.3}$$

In fact, setting $P(x) = \prod_{t=x_0}^{x-1} p(t)$, $P(x_0) = 1$ and dividing (1.5.3) by $P(x+1)$, we have

$$\frac{z(x+1)}{P(x+1)} - \frac{z(x)}{P(x)} = \frac{q(x)}{P(x+1)}.$$

If we write $y(x) = \dfrac{z(x)}{P(x)}$ and $g(x) = \dfrac{q(x)}{P(x+1)}$, equation (1.5.3) now takes the form $\Delta y(x) = g(x)$. The solution of (1.5.3) is then given by

$$\begin{aligned} z(x) &= P(x)\Delta^{-1}\frac{q(x)}{P(x+1)} + z_0 P(x) \\ &= P(x)\sum_{s=x_0}^{x-1}\frac{q(s)}{P(s+1)} + z_0 P(x) \\ &= \sum_{s=x_0}^{x-1} q(s)\prod_{t=s+1}^{x-1} p(t) + z_0 \prod_{t=x_0}^{x-1} p(t). \end{aligned} \tag{1.5.4}$$

Let us now consider the following nonlinear difference equation

$$y(x) = x\Delta y(x) + \phi(\Delta y(x)), \tag{1.5.5}$$

where ϕ is a nonlinear function. This difference equation is called the Clairaut equation in analogy with the similar differential equation of the corresponding name.

Letting $p(x) = \Delta y(x)$ and substituting in (1.5.5), we arrive at

$$y(x) = xp + \phi(p).$$

Applying Δ to both sides, the above relation becomes

$$p = p + (x+1)\Delta p + \phi(p + \Delta p) - \phi(p),$$

which reduces to $(x+1)\Delta p + \phi(p+\Delta p) - \phi(p) = 0$. This equation can be split into

$$\Delta p = 0, \tag{1.5.6}$$

and

$$\phi(p + \Delta p) - \phi(p) = -(x+1)\Delta p. \tag{1.5.7}$$

Hence, from (1.5.6), $p(x) = \omega(x)$, where $\omega(x)$ is an arbitrary function of period one and the corresponding solution of (1.5.5) is

$$y(x) = x\omega(x) + \phi(\omega(x)).$$

The solution of (1.5.7) gives rise to other solutions of (1.5.5) (see problem 1.23 for an example). In the continuous case, it is known that the solutions of the Clairaut equation can be obtained by solving two differential equations corresponding to (1.5.6) and (1.5.7) and that the solution of the second one is the envelope of the solutions of the first one. In the discrete case we cannot speak about envelopes. The solutions of (1.5.7) can have common points with the solutions of (1.5.6), because for equations in the form (1.5.5), the property of uniqueness of solutions may not be true.

Analogous to the differential equation of Riccati type, we have the difference equation

$$y(x)y(x+1) + p_1(x)y(x+1) + p_2(x)y(x) + p_3(x) = 0, \qquad (1.5.8)$$

where $p_1(x), p_2(x), p_3(x)$ are arbitrary functions defined on J_x^+.

Equation (1.5.8) can be reduced to a linear one by setting $y(x) = \dfrac{z(x+1)}{z(x)} - p_1(x)$. The resulting equation is

$$z(x+2) + [p_2(x) - p_1(x)]z(x+1) + [p_3(x) - p_1(x)p_2(x)]z(x) = 0. \qquad (1.5.9)$$

In the study of iterative processes (see Chapter 5) the following equation arises

$$y_{n+1} = \frac{\frac{1}{2}y_n^2}{1 - \sum_{j=0}^{n} y_n}. \qquad (1.5.10)$$

By setting $y_n = z_n - z_{n-1}$, $z_{-1} = 0$, equation (1.5.10) reduces to

$$z_{n+1} = z_n + \frac{\frac{1}{2}(z_n - z_{n-1})^2}{1 - z_n}, \qquad (1.5.11)$$

which is of second order. Now consider the first order difference equation

$$z_{n+1} = \frac{\frac{1}{2}z_n^2 - z_1}{z_n - 1}. \qquad (1.5.12)$$

Proposition 1.5.1. *The solution of* (1.5.12) *satisfies* (1.5.10).

Proof. By multiplying (1.5.12) by $z_n - 1$, we obtain

$$z_{n+1}z_n - z_{n+1} + \tfrac{1}{2}z_{n+1}^2 = \tfrac{1}{2}z_n^2 + \tfrac{1}{2}z_{n+1}^2 - z_1$$

from which it follows that $\frac{1}{2}z_{n+1}^2 - z_{n+1} + z_1 = \frac{1}{2}(z_{n+1} - z_n)^2$. Thus we have

$$z_{n+1} - z_n = \frac{\frac{1}{2}z_n^2 - z_n + z_1}{1 - z_n} = \frac{\frac{1}{2}(z_n - z_{n-1})^2}{1 - z_n}. \qquad \blacksquare$$

Equation (1.5.12) is said to be the first integral of (1.5.12). The solution of (1.5.12) can be written explicitly (see Problem 1.26).

1.6. Comparison Results

One of the most efficient methods of obtaining information on the behavior of solutions of difference equations, even when they cannot be solved explicitly, is the comparison principle. In general, the comparison principle is concerned with estimating a function satisfying a difference inequality by the solution of the corresponding difference equation. In this section, we shall present various forms of this principle.

Theorem 1.6.1. *Let $n \in N_{n_0}^+$, $r \geq 0$ and $g(n, r)$ be a function nondecreasing with respect to r for any fixed n. Suppose that for $n \geq n_0$, the inequalities,*

$$y_{n+1} \leq g(n, y_n), \tag{1.6.1}$$

$$u_{n+1} \geq g(n, u_n), \tag{1.6.2}$$

hold. Then

$$y_{n_0} \leq u_{n_0} \tag{1.6.3}$$

implies

$$y_n \leq u_n, \qquad \textit{for all } n \geq n_0. \tag{1.6.4}$$

Proof. Suppose that (1.6.4) is not true. Then, because of (1.6.3) there exists a $k \in N_{n_0}^+$ such that $y_k \leq u_k$ and $y_{k+1} > u_{k+1}$. It follows, using (1.6.1), (1.6.2) and the monotone character of g, that

$$g(k, u_k) \leq u_{k+1} < y_{k+1} \leq g(k, y_k) \leq g(k, u_k),$$

which is a contradiction. Hence the proof. $\blacksquare$

Usually in applications, (1.6.1) or (1.6.2) is an equation and the corresponding result is called the comparison principle.

Corollary 1.6.1. *Let $n \in N_{n_0}^+$, $k_n \geq 0$ and $y_{n+1} \leq k_n y_n + p_n$. Then, for $n \geq n_0$, we have*

$$y_n \leq y_{n_0} \prod_{s=n_0}^{n-1} k_s + \sum_{s=n_0}^{n-1} p_s \prod_{\tau=s+1}^{n-1} k_\tau. \tag{1.6.5}$$

Proof. Because $k_n \geq 0$ the hypotheses of Theorem 1.6.1 are verified. Hence $y_n \leq u_n$, where u_n is the solution of the linear difference equation

$$u_{n+1} = k_n u_n + p_n, \qquad u_{n_0} = y_{n_0}. \tag{1.6.6}$$

By (1.5.4), we see that the right-hand-side of (1.6.5) is the solution of (1.6.6). ■

Theorem 1.6.2. *Let $g(n, s, y)$ be defined on $N_{n_0}^+ \times N_{n_0}^+ \times \boldsymbol{R}$ and nondecreasing with respect to y. Suppose that for $n \in N_{n_0}^+$,*

$$y_n \leq \sum_{s=0}^{n-1} g(n, s, y_s) + p_n.$$

Then, $y_{n_0} \leq p_{n_0}$ implies $y_n \leq u_n$, $n \geq n_0$, where u_n is the solution of the difference equation

$$u_n = \sum_{s=0}^{n-1} g(n, s, u_s) + p_n, \; u_{n_0} = p_{n_0}.$$

Proof. If the claim is not true, then there exists a $k \in N_{n_0}^+$ such that $y_{k+1} > u_{k+1}$ and $y_s \leq u_s$ for $s \leq k$. But,

$$y_{k+1} - u_{k+1} \leq \sum_{s=0}^{k} [g(k+1, s, y_s) - g(k+1, s, u_s)] \leq 0$$

which is a contradiction. ■

Corollary 1.6.2. (*Discrete Gronwall inequality*). *Let $n \in N_{n_0}^+$, $k_n \geq 0$ and*

$$y_{n+1} \leq y_{n_0} + \sum_{s=n_0}^{n} [k_s y_s + p_s].$$

Then,

$$y_n \leq y_{n_0} \prod_{s=n_0}^{n-1} (1 + k_s) + \sum_{s=n_0}^{n-1} p_s \prod_{\tau=s+1}^{n-1} (1 + k_\tau)$$

$$\leq y_{n_0} \exp\left(\sum_{s=n_0}^{n-1} k_s\right) + \sum_{s=n_0}^{n-1} p_s \exp\left(\sum_{\tau=s+1}^{n-1} k_\tau\right), \; n \geq n_0.$$

Proof. The comparison equation is

$$u_n = u_{n_0} + \sum_{s=n_0}^{n-1} [k_s u_s + p_s],\ u_{n_0} = y_{n_0}.$$

This is equivalent to $\Delta u_n = k_n u_n + p_n$, the solution of which is

$$u_n = u_{n_0} \prod_{s=n_0}^{n-1} (1 + k_s) + \sum_{s=n_0}^{n-1} P_s \prod_{\tau=s+1}^{n-1} (1 + k_\tau).$$

The proof is complete by observing that $1 + k_s \leq \exp(k_s)$. ∎

We can prove the following Corollary in a similar way.

Corollary 1.6.3. *Let $n \in N^+_{n_0}$, $k_n \geq 0$ and*

$$y_{n+1} \leq P_{n+1} + \sum_{s=n_0}^{n} k_s y_s,\ y_{n_0} \leq P_{n_0}.$$

Then,

$$\begin{aligned} y_n &\leq P_{n_0} \prod_{s=n_0}^{n-1} (1 + k_s) + \sum_{s=n_0}^{n-1} q_s \prod_{\tau=s+1}^{n-1} (1 + k_\tau) \\ &\leq P_{n_0} \exp\left(\sum_{n=n_0}^{n-1} k_s \right) + \sum_{s=n_0}^{n-1} q_s \exp\left(\sum_{\tau=s+1}^{n-1} k_\tau \right), \end{aligned}$$

where $q_n = \Delta P_n$.

Another form of the foregoing result is as follows.

Corollary 1.6.4. *Let $n \in N^+_{n_0}$, $k_n > 0$ and*

$$y_{n+1} \leq P_{n+1} + \sum_{s=n_0}^{n} k_s y_s,\ y_{n_0} \leq P_{n_0}.$$

Then,

$$\begin{aligned} y_n &\leq P_n + \sum_{s=n_0}^{n-1} P_s k_s \sum_{\tau=s+1}^{n-1} (1 + k_\tau) \\ &\leq P_n + \sum_{s=n_0}^{n-1} P_s k_s \exp\left(\sum_{\tau=s+1}^{n-1} k_\tau \right). \end{aligned}$$

Proof. By setting

$$V_n = \sum_{s=n_0}^{n-1} k_s y_s,\ V_{n_0} = 0. \tag{1.6.7}$$

we have

$$y_n \le P_n + V_n. \tag{1.6.8}$$

Applying the operator Δ to both sides of (1.6.7), we get

$$\Delta V_n = k_n y_n \le k_n P_n + k_n V_n$$

from which by Corollary 1.6.1, it follows that

$$V_n \le \sum_{s=n_0}^{n-1} k_s p_s \prod_{\tau=s+1}^{n-1} (1 + k_\tau).$$

In view of (1.6.8), we obtain the desired estimate. ■

Corollary 1.6.5. *Let $k(n, s, x)$: $N_{n_0}^+ \times N_{n_0}^+ \to R^+ \to R^+$ be monotonic nondecreasing in x and $g(n, u): N_{n_0}^+ \times R^+ \to R^+$ be monotonic nondecreasing in u. Suppose that*

$$y_n \le g\left(n, \sum_{s=n_0}^{n-1} k(n, s, y_s)\right). \tag{1.6.9}$$

Then

$$y_n \le g(n, u_n),$$

where u_n is the solution of

$$u_n = \sum_{s=n_0}^{n-1} k(n, s, g(s, u_s)).$$

Proof. The relation (1.6.9) can be written as $y_n \le g(n, r_n)$, where $r_n = \sum_{s=n_0}^{n-1} k(n, s, y_s)$. But this yields $r_n \le \sum_{s=n_0}^{n-1} k(n, s, g(s, r_s))$. Now applying Theorem 1.6.2, we obtain $r_n \le u_n$, which in turn shows

$$y_n \le g(n, r_n) \le g(n, u_n). \quad ■$$

Theorem 1.6.3. *Suppose that $g_1(n, u)$ and $g_2(n, u)$ are two functions defined on $N_{n_0}^+ \times R^+$ and nondecreasing with respect to u. Let*

$$g_2(n, u_n) \le u_{n+1} \le g_1(n, u_n).$$

Then,

$$P_n \le u_n \le r_n,$$

where P_n and r_n are the solutions of the difference equations:

$$r_{n+1} = g_1(n, r_n), \qquad r_0 \le u_0,$$

$$P_{n+1} = g_2(n, P_n), \qquad P_0 \ge u_0.$$

Proof. Applying Theorem 1.6.1 twice, we obtain the needed estimate. ■

Theorem 1.6.4. (*Discrete Bihari inequality*). *Suppose that h_n is a nonnegative function defined on $\mathbf{N}_{n_0}^+$, $M > 0$ and W is a positive strictly increasing function defined on $\mathbf{R}^+$. If for $n \geq n_0$, $y_n \leq V_n$ where $V_n = y_0 + M\sum_{s=n_0}^{n-1} h_s W(y_s)$, then, for $n \in \mathbf{N}_1$,*

$$y_n \leq G^{-1}\left(G(y_0) + M\sum_{s=n_0}^{n-1} h_s\right),$$

where G is the solution of

$$\Delta G(V_n) = \frac{\Delta V_n}{W(V_n)}$$

and $N_1 = \{n \in \mathbf{N}_{n_0}^+ \mid M\sum_{s=n_0}^{n-1} h_s \leq G(\infty) - G(x_0)\}$.

Proof. We have $\Delta V(n) = Mh_n W(y_n) \leq Mh_n W(V_n)$. It follows that

$$G(V_{n+1}) \leq G(V_n) + Mh_n,$$

from which, in view of Theorem 1.6.1, we get

$$G(V_n) \leq G(y_0) + M\sum_{s=n_0}^{n-1} h_s.$$

Hence, for $n \in \mathbf{N}_1$, $V_n \leq G^{-1}(G(x_0) + M\sum_{s=n_0}^{n-1} h_s)$. ■

Theorem 1.6.5. *Let a_n, b_n be two nonnegative functions defined on $\mathbf{N}_{n_0}^+$ and P_n be a positive nondecreasing function defined on $\mathbf{N}_{n_0}$. Let*

$$y_n \leq P_n + \sum_{s=n_0}^{n-1} a_s y_s + \sum_{s=n_0}^{n-1} a_s\left(\sum_{k=n_0}^{n-1} b_k y_k\right).$$

Then,

$$y_n \leq P_n\left[1 + \sum_{s=n_0}^{n-1} a_s \prod_{k=n_0}^{s-1}(1 + a_k + b_k)\right].$$

Proof. Since P_s is positive and nondecreasing, we obtain

$$\frac{y_n}{P_n} \leq 1 + \sum_{s=n_0}^{n-1} a_s\frac{y_s}{P_s} + \sum_{s=n_0}^{n-1} a_s\left(\sum_{k=n_0}^{n-1} b_k\frac{y_k}{P_k}\right).$$

We let

$$u_n = 1 + \sum_{s=n_0}^{n-1} a_s\frac{y_s}{P_s} + \sum_{s=n_0}^{n-1} a_s\left(\sum_{k=n_0}^{n-1} b_k\frac{y_k}{P_k}\right), \qquad u_{n_0} = 1,$$

so that we have

$$\Delta u_n = a_n\left[\frac{y_n}{P_n} + \sum_{k=n_0}^{n-1} b_k \frac{y_k}{P_k}\right].$$

This in turn yields

$$\Delta u_n \leq a_n\left[u_n + \sum_{k=n_0}^{n-1} b_k u_k\right].$$

By setting $x_n = u_n + \sum_{k=n_0}^{n-1} b_k u_k$, we obtain

$$\Delta x_n = \Delta u_n + b_n u_n \leq a_n x_n + b_n u_n \leq (a_n + b_n)x_n,$$

from which one gets $x_{n+1} \leq (1 + a_n + b_n)x_n$, and therefore,

$$x_n \leq \prod_{k=n_0}^{n-1} (1 + a_k + b_k).$$

Consequently, we arrive at

$$\Delta u_n \leq a_n \prod_{k=n_0}^{n-1} (1 + a_k + b_k),$$

which implies

$$u_n \leq 1 + \sum_{s=n_0}^{n-1} a_s \sum_{k=n_0}^{n-1} (1 + a_k + b_k).$$

This in turn yields $y_n \leq P_n u_n$. ∎

Theorem 1.6.6. *Let $y_j (j = 0, 1, \ldots)$ be a positive sequence satisfying*

$$y_{n+1} \leq g\left(y_n, \sum_{j=0}^{n-1} y_j, \quad \sum_{j=0}^{n-2} y_j\right),$$

where $g(y, z, w)$ is a nondecreasing function with respect to its arguments. If $y_0 \leq u_0$ and

$$u_{n+1} = g\left(u_n, \sum_{j=0}^{n-1} u_j, \quad \sum_{j=0}^{n-2} u_j\right),$$

then, for all $n \geq 0$

$$y_n \leq u_n.$$

Proof. The proof is by induction. The claim is true for $n = 0$. Suppose it is true for $n = k$. Then, we have

$$y_{k+1} \leq g\left(y_k, \sum_{j=0}^{k-1} y_j, \quad \sum_{j=0}^{k-2} y_j\right) \leq g\left(u_k, \sum_{j=0}^{k-1} u_j, \quad \sum_{j=0}^{k-2} u_j\right) = u_{k+1}. \quad ∎$$

If we take $u_k = t_{k+1} - t_k = \Delta t_k$, $t_0 = 0$, the comparison equation becomes

$$\Delta t_k = g(\Delta t_k, t_k, t_{k-1}).$$

Theorem 1.6.7. *Let y_i $(i = 0, 1, \ldots)$ be a positive sequence satisfying the inequality*

$$y_{n+1} \leq g(y_n, y_{n-1}, \ldots, y_{n-k}),$$

where g is nondecreasing with respect to its arguments. Then,

$$y_n \leq u_n,$$

where u_n is the solution of

$$u_{n+1} = g(u_n, u_{n-1}, \ldots, u_{n-k}), \qquad y_j \leq u_j, \qquad j = 0, 1, \ldots, k.$$

Proof. Suppose that the conclusion is not true. Then there exists an index $m \geq k$ such that $y_{m+1} > u_{m+1}$ and $y_j \leq u_j$, $j \leq m$. It follows that

$$g(y_m, y_{m-1}, \ldots, y_{m-k}) \geq y_{m+1} > u_{m+1} = g(u_m, u_{m-1}, \ldots, u_{m-k}),$$

which is a contradiction. ■

The next theorem deals with the higher difference of the unknown function.

Theorem 1.6.8. *Let y_i, p_i, q_i be nonnegative sequences defined on $\mathbf{N}_0^+$ and h_{ij} be a nonnegative sequence defined on $\mathbf{N}^+ \times \mathbf{N}^+$. Suppose that, for $k \geq 0$,*

$$\Delta^k u_n \leq p_n + q_n \sum_{j=0}^{k} \sum_{s=0}^{n-1} h_{js} \Delta^j u_s. \tag{1.6.10}$$

Then,

$$\Delta^k u_n \leq p_n + q_n \sum_{s=0}^{n-1} \phi_s \sum_{\tau=s+1}^{n-1} [1 + \psi_\tau],$$

where

$$\phi_s = p_n h_{kn} + \sum_{j=0}^{k-1} \sum_{i=0}^{j} (\Delta^j u_0) h_{in} \binom{n}{j-i}$$
$$+ \sum_{j=0}^{k-1} h_{k-j-1,n} \sum_{s=0}^{n-j-1} \binom{n-s-1}{j} p_s,$$
$$\psi_n = q_n h_{kn} + \sum_{j=0}^{k-1} h_{k-j-1,n} \sum_{s=0}^{n-j-1} \binom{n-s-1}{j} q_s.$$

Proof. From the discrete Taylor formula (1.3.5), we have, for $0 \leq j \leq k-1$,

$$\Delta^j u_n = \sum_{i=j}^{k-1} \binom{n}{i-j} \Delta^i u_0 + \sum_{s=0}^{n-k+j} \binom{n-s-1}{k-j-1} \Delta^k u_s. \tag{1.6.11}$$

Define $R_n = \sum_{j=0}^{k} \sum_{s=0}^{n-1} h_{js} \Delta^j u_s$, and $R_0 = 0$. The inequality (1.6.10) can be written as

$$\Delta^k u_n \leq p_n + q_n R_n. \tag{1.6.12}$$

From the definition of R_n, one gets

$$\Delta R_n = \sum_{j=0}^{k} h_{jn} \Delta^j u_n = h_{kn} \Delta^k u_n + \sum_{j=0}^{k-1} h_{jn} \Delta^j u_n,$$

and by using (1.6.11) and (1.6.12), it follows that

$$\Delta R_n \leq h_{kn}(p_n + q_n R_n) + \sum_{j=0}^{k-1} h_{jn} \sum_{i=j}^{k-1} \binom{n}{i-j} \Delta^i u_0$$
$$+ \sum_{j=0}^{k-1} h_{jn} \sum_{s=0}^{n-k+j} \binom{n-s-1}{k-j-1} \Delta^k u_s.$$

By exchanging the indices i and j, reversing the summations in the second term and changing indices in the third expression, we obtain

$$\Delta R_n \leq h_{kn}(p_n + q_n R_n) + \sum_{j=0}^{k-1} \sum_{i=0}^{j} \binom{n}{j-i} (\Delta^j u_0) h_{in}$$
$$+ \sum_{j=0}^{k-1} h_{k-j-1} \sum_{s=0}^{n-j-1} \binom{n-s-1}{j} (p_s + q_s R_s)$$
$$\leq h_{kn} p_n + \sum_{j=0}^{k-1} \sum_{i=0}^{j} \binom{n}{j-i} (\Delta^j u_0) h_{in} + \sum_{j=0}^{k-1} h_{k-j-1,n} \sum_{s=0}^{n-j-1} \binom{n-s-1}{j} p_s$$
$$+ \left[h_{kn} q_n + \sum_{j=0}^{k-1} h_{k-j-1,n} \sum_{s=0}^{n-j-1} \binom{n-s-1}{j} q_s \right] R_n,$$

where the nondecreasing property of R_s has been used. By the definitions of the functions ϕ_n and ψ_n, we get

$$\Delta R_n \leq \phi_n + \psi_n R_n.$$

The proof is complete by using Corollary 1.6.1, which gives

$$R_n \leq \sum_{s=0}^{n-1} \phi_s \prod_{\tau=s+1}^{n-1} (1 + \psi_\tau),$$

and from (1.6.12), it follows that

$$\Delta^k u_n \leq p_n + q_n \sum_{s=0}^{n-1} \phi_s \prod_{\tau=s+1}^{n-1} (1 + \psi_\tau). \qquad \blacksquare$$

1.7. Problems

1.1. Let $y_1 = -2$, $y_2 = -2$, $y_3 = 0$, $y_4 = 4$. Compute y_5 supposing that $\Delta^4 y_1 = 0$.

1.2. Show that

$$\Delta\, e^{i(ax+b)} = e^{i(ax+b)}(e^{ia} - 1),$$

$$\Delta^{-1}\, e^{i(ax+b)} = \frac{e^{i(ax+b)}}{e^{ia} - 1} + \omega(x).$$

1.3. Using the result of problem 1.2, show that

$$\Delta \cos(ax + b) = -2 \sin \frac{a}{2} \sin\left(ax + b + \frac{a}{2}\right),$$

$$\Delta \sin(ax + b) = 2 \sin \frac{a}{2} \cos\left(ax + b + \frac{a}{2}\right).$$

1.4. Show that $\sum_{j=1}^{n} j^3 = \frac{1}{4} n^2 (n+1)^2$.

1.5. Prove that $\sum_{i=1}^{n} \cos qi = \dfrac{\sin(n + \frac{1}{2})q - \sin \frac{1}{2}q}{2 \sin \frac{1}{2}q}$.

1.6. The factorial powers can be defined for values of x that are not integers. Letting $m = 1$ and $n = x - 1$, in the relation $x^{(m+m)} = x^{(m)}(x - m)^{(n)}$, we get $x^{(x)} = x(x-1)^{(x-1)}$. Setting $y(x) = x^{(x)}$, the previous expression can be written as $y(x) = xy(x-1)$. The solution of this equation is $y(x) = \Gamma(x+1)$, where $\Gamma(x)$ is the Euler function, which is defined for all real values of x except the negative integers. This allows us to define $x^{(x)}$ for $x = 0$. In fact, we have $0^{(0)} = \Gamma(1) = 1$. From $x^{(-n)} = \dfrac{1}{(x+1)(x+2)\ldots(x+n)}$ we also have $0^{(-n)} = \dfrac{1}{n!}$.

1.7. Using the previous result, show that $x^{(n)} = \dfrac{\Gamma(x+1)}{\Gamma(x+1-n)}$.

1.8. Show that $\Delta \log \Gamma(x) = \log x$ and $\Delta^{-1} \log x = \log \Gamma(x) + \omega(x)$.

1.9. Let $\psi(x) = \dfrac{\Gamma'(x)}{\Gamma(x)}$, where $\Gamma'(x)$ is the derivative of $\Gamma(x)$. Show that $\Delta \psi(x) = \dfrac{1}{x}$. The function $\psi(x)$ has many interesting properties. Among them we recall $\lim_{x \to \infty} (\psi(x) - \log(x)) = 0$.

1.10. Show that, if $p(x)$ is a degree k polynomial, it can be written in the form $p(x) = \sum_{i=0}^{k} \frac{x^{(i)}}{i!}\Delta^i p(0)$, which is the Newton form.

1.11. Show that for $q, n \in \mathbf{N}_0^+$, $\sum_{j=0}^{q-n} (-1)^j \binom{q}{n+j} = \binom{q-1}{n-1}$. (Hint: Use the Sterling formula $\binom{q}{n+j} = \binom{q-1}{n+j} + \binom{q-1}{n+j-1}$.

1.12. Show that for $q, l, n \in \mathbf{N}_0^+$, one has

$$S(q, l, n) \equiv \sum_{j=l}^{q-n} (-1)^{j-l} \binom{q}{n+j}\binom{j}{l} = \binom{q-l-1}{n-1}.$$

Note that $S(q, 0, n) = \binom{q-1}{n-1}$ as established in the previous exercise.

1.13. Verify that, if $\{y_n\}$ and $\{z_n\}$ are two sequences, one has

(a) $\Delta(y_n z_n) = y_{n+1}\Delta z_n + z_n \Delta y_n$

$\qquad = y_n \Delta z_n + z_{n+1}\Delta y_n$

$\qquad = y_n \Delta z_n + z_n \Delta y_n + \Delta y_n \Delta z_n,$

(b) $\Delta \frac{y_n}{z_n} = \frac{z_n \Delta y_n - y_n \Delta z_n}{z_n z_{n+1}},$

(c) $\sum_{n=0}^{N-1} y_n \Delta z_n = y_n z_n \Big|_{n=0}^{n=N} - \sum_{n=0}^{N-1} z_{n+1}\Delta y_n,$

the last formula is called summation by parts.
By setting $\Delta z_n = b_n$ we have,

(d) $\sum_{n=0}^{N-1} y_n b_n = y_N \sum_{i=0}^{N} b_i - \sum_{n=0}^{N-1} (\sum_{i=0}^{n} b_i)\Delta y_n$, known as the Abel formula.

1.14. Let $f(x) \in C^\infty[a, b]$. Putting $E_h f(x) = f(x + h)$, with $x, x + h \in [a, b]$, prove that $E_h = e^{hD} = I + \Delta_h$, where $\Delta_h f(x) = f(x + h) - f(x)$ and $Df(x) = \frac{df(x)}{dx}$.

1.15. Expand in power series the function $f(x) = \frac{x}{e^x - 1}$ and verify that $f(x) = \sum_{n=0}^{\infty} \frac{B_n}{n!} x^n$, where B_n are the Bernoulli numbers.

1.16. Prove that $\frac{x}{e^x - 1} + \frac{x}{2} = \frac{x \cosh x/2}{2 \sinh x/2}$ and using the result of problem 1.13, show that $B_{2k+1} = 0$, $k = 1, 2, \ldots$.

1.17. Using the result of problem 1.12, show that $\Delta^{-1} = \sum_{n=0}^{\infty} \frac{B_n}{n!} D^{n-1}$.

1.18. (*Euler–McLaurin formula*). From the previous results, show that

$$D^{-1}f(x) = \int f(x)\, dx = \Delta^{-1}f(x) - \sum_{n=1}^{\infty} \frac{B_n}{n!} D^{n-1}f(x).$$

This formula can be used either to approximate an integral or to approximate a sum using an integral.

1.19. Using the result of problem 1.14 prove that $D^{-1} = \Delta^{-1} + \frac{1}{2}I - \frac{1}{12}\Delta + \frac{1}{24}\Delta^2 - \ldots$, from which one obtains the Newton integration formula

$$\int f(x)\, dx = \Delta^{-1}f(x) + \tfrac{1}{2}f(x) - \tfrac{1}{12}\Delta f(x) + \tfrac{1}{24}\Delta^2 f(x) \ldots.$$

1.20. Show that

$$D = \frac{1}{h}\left[\Delta - \frac{1}{2}\Delta^2 + \frac{1}{3}\Delta^3 \ldots\right] = \frac{1}{h}\sum_{i=1}^{\infty} (-1)^i \frac{\Delta^i}{i}.$$

This formula can be used to approximate numerically the derivative of a function.

1.21. (*Horner rule*). Find the solution of the difference equation

$$y_n = z y_{n-1} + b_n, \qquad y_0 = b_0.$$

1.22. Find the solution of $y_{n+1} + y_n = n + 1$.

1.23. Solve the Clairaut equation $y = n\Delta y - \phi(\Delta y)$ and verify that for $n = 0$ there are two solutions.

1.24. Solve the following nonlinear equation $y_{n+1}(1 + y_n) = y_n$.

1.25. In many applications (for example in computing the cost of Fast Fourier transform algorithm) one meets the following equation $C(n) = 2C\left(\frac{n}{2}\right) + f(n)$, $C(2) = 2$. Transform this equation into a linear one and find the solution.

1.26. The Newton method to compute the square root of a positive number A reduces to the difference equation $z_{n+1} = \frac{1}{2}\left(z_n + \frac{A}{z_n}\right)$. Transform this equation to a linear one and find the solution.

1.27. Solve the following nonlinear equations

(A) $y_{n+1} = y_n^2$, (B) $y_{n+1} = 2y_n(1 - y_n)$, (C) $y_{n+1} = \dfrac{z_0 - \frac{1}{2}y_n^2}{1 - y_n}$

1.28. Find the first integral of

$$y_{n+2} = y_n \frac{(1 + y_n y_{n+1})}{1 + y_{n+1}^2}.$$

1.29. Suppose that $c \geq 0$; $k_n \geq 0$, $k_n < \varepsilon$ for $n \geq n_0$ and $y_n \leq c + \sum_{s=n_0}^{n} k_s y_s$. Show that $y_n \leq \dfrac{c}{1 - \varepsilon} \exp\left(\dfrac{1}{1 - \varepsilon} \sum_{s=n_0}^{n-1} k_s\right)$.

1.30. Let be $h > 0$, $M > 0$, $\delta > 0$ and $u_n \leq h^{1/2} M \sum_{j=0}^{n-1} \dfrac{1}{(n - j)^{1/2}} u_j + \delta$, $0 < u_0 \leq \delta$. Find a bound for u_n.

1.31. Suppose that $c \geq 0$, $k_n \geq 0$ for $n \geq n_0$ and $y_n \leq c + \sum_{s=n+1}^{\infty} k_s y_s$. Show that $y_n \leqq c \exp(\sum_{s=n+1}^{\infty} k_s)$.

1.32. Study the following inequality $y_{n+1}^2 \leq a y_n^2 + b_n(y_n + y_{n+1})$ for $a > 1$ and $a < 1$.

1.8. Notes

Most of the contents of Sections 1.1 to 1.4 may be found in many classical books on difference equations, such as Jordan [83], Milne and Thomson [116], Fort [49], Brand [16] and Miller [114]. Propositions 1.3.4 and 1.3.5 have been adapted from Agarwal [5]. Condition 1.4.8 can be found in Schoenberg [153]. Theorem 1.6.1 and many other comparison results on difference inequalities are taken from Sugiyama [163]. Theorem 1.6.2 was established by Maslovskaya [13]. Theorems 1.6.3, 1.6.4 and 1.6.5 are due to Pachpatte [133, 134, 135]. Theorem 1.6.6 is adapted from a result of Rheinboldt [147]. Theorem 1.6.7 is essentially new. Theorem 1.6.8 is due to Agarwal and Thandapani [1]. A generalization of this theorem can be found in Agarwal and Thandapani [1, 2]. For allied results on difference inequalities, see Popenda [142, 143], Mate' and Nevai [101] and Patula [139].

CHAPTER 2 Linear Difference Equations

2.0. Introduction

This chapter investigates the essential techniques employed in the treatment of linear difference equations. We begin Section 2.1 with the fundamental theory of linear difference equations and develop the method of variation of constants in Section 2.2. We then specialize our discussions to linear difference equations with constant coefficients, since this is an important class in itself. We exhibit, in Sections 2.3 to 2.5, methods of obtaining solutions of such equations by the use of difference operators as well as generating functions that offer elegant methods of solving difference equations. In Section 2.6, we present theory of stability for difference equations with constant coefficients and discuss in Section 2.7 linear multistep methods as an application of the theory of absolute stability. Section 2.8 considers the scalar boundary value problem that is needed later on. Section 2.9 completes the treatment by filling in the gaps with interesting problems.

2.1. Fundamental Theory

Since it is not essential to specify the properties of functions with respect to $x \in \mathbb{C}$, we shall use in this section $\boldsymbol{N}^+$ or $\boldsymbol{N}^+_{n_0}$ as definition set. We shall denote a sequence by $\{y_n\}$, which is the set of all values of the function y on $\boldsymbol{N}^+_{n_0}$.

Definition 2.1.1. Let $p_0(n) = 1$, $p_1(n), \ldots, p_k(n)$, g_n be $k+2$ functions defined on $N_{n_0}^+$. An equation of the form

$$y_{n+k} + p_1(n)y_{n+k-1} + \ldots + p_k(n)y_n = g_n \tag{2.1.1}$$

is called a linear difference equation of order k, provided that $p_k(n) \neq 0$.

With (2.1.1), the following k initial conditions

$$y_{n_0} = c_1, \qquad y_{n_0+1} = c_2, \ldots, y_{n_0+k-1} = c_k \tag{2.1.2}$$

are associated, where c_i are real or complex constants. For convenience, we shall state the following theorem, which is a special case of Theorem 1.5.1.

Theorem 2.1.1. *The equation* (2.1.1) *with the initial conditions* (2.1.2) *has a unique solution.* ■

We shall denote by $y(n, n_0, c)$ the solution of (2.1.1) and (2.1.2), where $c = (c_1, c_2, \ldots, c_k) \in \mathbb{C}^k$. Thus, we have

$$y(n_0 + j, n_0, c) = c_{j+1}, \qquad j = 0, 1, \ldots, k-1.$$

Definition 2.1.2. If for every $n \in N_{n_0}^+$, $g(n) = 0$, then the equation (2.1.1) is said to be homogeneous.

Introducing the operator L defined by

$$Ly_n = \sum_{i=0}^{k} p_i(n)y_{n+k-i} \tag{2.1.3}$$

equation (2.1.1) assumes the form

$$Ly_n = g(n) \tag{2.1.4}$$

and the homogeneous one becomes

$$Ly_n = 0. \tag{2.1.5}$$

It is easy to verify that L is linear, since

$$L(\alpha y_n + \beta z_n) = \alpha L y_n + \beta L z_n,$$

when $\alpha, \beta \in C$ and $\{y_n\}$, $\{z_n\}$ are two sequences.

Let S be the space of solutions of (2.1.5). Because of linearity of L, we have the following result.

Lemma 2.1.1. *Any linear combination of elements of S lies in S. Let $y(n, n_0, E_1), \ldots, y(n, n_0, E_k)$ be solutions of* (2.1.5), *where*

$$E_1 = (1, 0, \ldots, 0), \qquad E_2 = (0, 1, 0, \ldots, 0), \ldots, E_k = (0, 0, \ldots, 0, 1),$$

and therefore,

$$Ly(n, n_0, E_i) = 0, \qquad i = 1, 2, \ldots, k. \tag{2.1.6}$$

Lemma 2.1.2. *Any other solution $y(n, n_0, c)$ of (2.1.5) can be expressed as a linear combination of $y(n, n_0, E_i)$, $i = 1, 2, \ldots, k$.*

Proof. If $y(n, n_0, c)$ is the solution with initial conditions $c_1, c_2, \ldots, c_k$, then the sum

$$z_n = \sum_{i=1}^{k} c_i y(n, n_0, E_i) \tag{2.1.7}$$

is a solution of (2.1.5) by Lemma 2.1.1 and also satisfies the same initial conditions $z_{n_0} = c_1$, $z_{n_0+1} = c_2, \ldots, z_{n_0+k-1} = c_k$. By Theorem 2.1.1 the two solutions must coincide. ∎

Let us now take k functions $f_i(n)$ defined on $N_{n_0}^+$ and define the matrix

$$K(n) = \begin{pmatrix} f_1(n) & f_2(n) & \ldots & f_k(n) \\ f_1(n+1) & f_2(n+1) & \ldots & f_k(n+1) \\ \ldots & \ldots & \ldots & \ldots \\ \ldots & \ldots & \ldots & \ldots \\ f_1(n+k-1) & f_2(n+k-1) & \ldots & f_k(n+k-1) \end{pmatrix}.$$

Definition 2.1.3. The function $f_i(n)$, $i = 1, 2, \ldots, k$ are linearly independent if for all n,

$$\sum_{i=1}^{k} \alpha_i f_i(n) = 0 \tag{2.1.8}$$

implies $\alpha_i = 0$, $i = 1, 2, \ldots, k$.

Theorem 2.1.2. *A sufficient condition for the functions $f_i(n)$, $i = 1, 2, \ldots, k$, to be linearly independent is that there exists an $\bar{n} \geq n_0$ such that $\det K(\bar{n}) \neq 0$.*

Proof. If (2.1.8) holds, then

$$\begin{aligned} &\sum_{i=1}^{k} \alpha_i f_i(n) = 0, \\ &\sum_{i=1}^{k} \alpha_i f_i(n+1) = 0, \\ &\quad \vdots \qquad \vdots \\ &\sum_{i=1}^{k} \alpha_i f_i(n+k-1) = 0. \end{aligned} \tag{2.1.9}$$

This linear homogeneous system of k equations in k unknowns has the coefficient matrix $K(n)$. Thus, if for $n = \bar{n}$, $\det K(\bar{n}) \neq 0$, the unique solution of the system is $\alpha_i = 0$, $i = 1, \ldots, k$, and the functions are linearly independent. ■

Theorem 2.1.3. *If* $f_i(n)$, $i = 1, 2, \ldots, k$, *are solutions of* (2.1.5), *then* $\det K(n) \neq 0$, *for* $n \geq n_0$, *provided that* $\det K(n_0) \neq 0$.

Proof. From the definition of the matrix $k(n)$, we get

$$\det K(n_0+1) = \det\begin{pmatrix} f_1(n_0+1) & f_2(n_0+1) & \ldots & f_k(n_0+1) \\ f_1(n_0+2) & f_2(n_0+2) & \ldots & f_k(n_0+2) \\ \ldots & \ldots & \ldots & \ldots \\ f_1(n_0+k) & f_2(n_0+k) & \ldots & f_k(n_0+k) \end{pmatrix}$$

$$= -p_k(n_0)\det\begin{pmatrix} f_1(n_0+1) & \ldots & f_k(n_0+1) \\ f_1(n_0+2) & \ldots & f_k(n_0+1) \\ \ldots & \ldots & \ldots \\ \ldots & \ldots & \ldots \\ f_1(n_0+k-1) & \ldots & f_k(n_0+k-1) \\ f_1(n_0) & \ldots & f_k(n_0) \end{pmatrix}$$

$$= (-1)^k p_k(n_0)\det K(n_0).$$

Since $p_k(n_0) \neq 0$, it follows that $\det K(n_0+1) \neq 0$. In general, for every $n \in N^+_{n_0}$, we have

$$\det K(n) = (-1)^k p_k(n) \det K(n-1), \tag{2.1.10}$$

and thus by induction it is now easy to see that $\det K(n) \neq 0$. ■

Corollary 2.1.1. *The solutions* $y(n, n_0, E_i)$, $i = 1, 2, \ldots, k$, *are linearly independent.*

Proof. In this case, $\det K(n_0) = 1$, since

$$K(n_0) = \begin{pmatrix} 1 & 0 & \ldots & 0 \\ 0 & \ddots & \ddots & \vdots \\ \vdots & \ddots & \ddots & 0 \\ 0 & \ldots & 0 & 1 \end{pmatrix}. \quad ■$$

In view of Corollary 2.1.1, the base $y(n, n_0, E_i)$, $i = 1, 2, \ldots, k$, is said to be a canonical base. Using Lemma 2.1.2 and Corollary 2.1.1, we have the following result.

Theorem 2.1.4. *The space S of solutions of* (2.1.5) *is a vector space of dimension k.*

If $a_i \in \mathbb{C}^k$, $i = 1, 2, \ldots, k$, are linearly independent, the set of solutions $y(n, n_0, a_i)$ can also be used as a base of the space S. In fact, in this case $K(n_0) = (a_1, \ldots, a_k)$ where a_i is the ith column of the matrix $K(n_0)$ and because they are linearly independent, it follows that $\det K(n) \neq 0$ for all $n \in \mathbf{N}_{n_0}^+$.

The matrix $K(n)$ is called the matrix of Casorati and it plays in the theory of difference equations the same role as the Wronskian matrix in the theory of linear differential equations.

Definition 2.1.4. Given k linearly independent solutions of (2.1.5), any linear combination of them is said to be a general solution of (2.1.5).

The term, general, means that such a solution can satisfy any set of initial conditions.

Lemma 2.1.3. *The difference between two solutions* y_n *and* $\bar{y}_n$ *of* (2.1.4) *satisfies* (2.1.5).

Proof. One has $Ly_n = g_n$ and $L\bar{y}_n = g_n$, which implies $L(y_n - \bar{y}_n) = 0$. ■

Theorem 2.1.5. *Let* $y(n, n_0, a_i)$, $i = 1, 2, \ldots, k$, *be* k *linearly independent solutions of* (2.1.5) *and* $\bar{y}_n$ *be a solution of* (2.1.4). *Then, any other solution of* (2.1.4) *can be written as*

$$y_n = \bar{y}_n + \sum_{i=1}^{k} a_i y(n, n_0, a_i). \tag{2.1.11}$$

Proof. From Lemma 2.1.3, one has that $y_n - \bar{y}_n \in S$ and therefore, it can be expressed as a linear combination of $y(n, n_0, a_i)$. ■

The foregoing theorem also means that the general solution of (2.1.4) is obtained by adding to the general solution of (2.1.5) any solution of (2.1.4).

Associated with the operator L, one defines the adjoint operator L^* by $L^* y_n = \sum_{i=0}^{k} p_i(n+i) y_{n+i}$ and the adjoint equations

$$L^* y_n = 0, \tag{2.1.12}$$

and

$$L^* y_n = b_n. \tag{2.1.13}$$

There are some interesting properties connecting the solutions of the equations and its adjoint. We shall consider, however, because of its use in Numerical Analysis, (see section 5.4) the transpose operator L^T defined by

$$L^T y_n = \sum_{i=0}^{k} p_i(n-k+i) y_{n+i} \tag{2.1.14}$$

and the transpose equation

$$L^T y_n = g_n. \tag{2.1.15}$$

Theorem 2.1.8. *Let u_n be the solution of the homogeneous equation* (2.1.5) *and y_n the solution of* (2.1.15) *with $y_{N+j} = 0$ for $j = 1, 2, \ldots, k$ and $N > k$. Then*

$$\sum_{n=0}^{N} u_n g_n = \sum_{j=0}^{k-1} y_j \sum_{i=0}^{j} p_i(j-k) u_{j-i}$$

Proof. We have

$$\sum_{n=0}^{N} u_n g_n = \sum_{n=0}^{N} u_n L^T y_n = \sum_{n=0}^{N} \sum_{i=0}^{k} u_n p_i(n-k+i) y_{n+i}.$$

Setting $j = n + i$ and $s = \min(j, k)$, one has

$$\sum_{n=0}^{N} u_n g_n = \sum_{j=0}^{N+k} y_j \sum_{i=0}^{s} p_i(j-k) u_{j-i}, \qquad p_i \equiv 0 \quad \text{for} \quad i > k.$$

Since $y_j = 0$ for $j > N$ and $\sum_{i=0}^{s} p_i(j-k) u_{j-i} = 0$ for $s \geq k$, the conclusion follows immediately. ∎

2.2. The Method of Variation of Constants

It is possible to find a particular solution of (2.1.4) knowing the general solution of (2.1.5). This may be accomplished by the method of variation of constants. Let $y(n, n_0, c)$ be a solution of (2.1.5) and $y(n, n_0, E_i)$, $i = 1, 2, \ldots, k$, be the canonical base in the space S of solutions of (2.1.5). Then,

$$y(n, n_0, c) = \sum_{j=1}^{k} c_j y(n, n_0, E_j). \tag{2.2.1}$$

We shall consider now the c_j as functions of n with

$$c_j(n_0) = c_j \tag{2.2.2}$$

and require that the functions

$$y(n, n_0, c(n)) = \sum_{j=1}^{k} c_j(n) y(n, n_0, E_j) \tag{2.2.3}$$

satisfies the equation (2.1.4). From (2.2.3) it now follows that

$$\begin{aligned} y(n+1, n_0, c(n+1)) &= \sum_{j=1}^{k} c_j(n+1)y(n+1, n_0, E_j) \\ &= \sum_{j=1}^{k} c_j(n)y(n+1, n_0, E_j) \\ &\quad + \sum_{j=1}^{k} \Delta c_j(n)y(n+1, n_0, E_j). \end{aligned}$$

By setting

$$\sum_{j=1}^{k} \Delta c_j(n)y(n+1, n_0, E_j) = 0, \tag{2.2.4}$$

we have

$$y(n+1, n_0, c(n+1)) = \sum_{j=1}^{k} c_j(n)y(n+1, n_0, E_j).$$

Similarly, for $i = 1, 2, \ldots, k-1$, we can get

$$y(n+i, n_0, c(n+i)) = \sum_{j=1}^{k} c_j(n)y(n+i, n_0, E_j), \tag{2.2.5}$$

if we set

$$\sum_{j=1}^{k} \Delta c_j(n)y(n+i, n_0, E_j) = 0, \qquad i = 1, 2, \ldots, k-1. \tag{2.2.6}$$

Therefore, in the end, we obtain

$$\begin{aligned} y(n+k, n_0, c(n+k)) &= \sum_{j=1}^{k} c_j(n)y(n+k, n_0, E_j) \\ &\quad + \sum_{j=1}^{k} \Delta c_j(n)y(n+k, n_0, E_j). \end{aligned}$$

By substituting in (2.14), there results

$$\begin{aligned} &\sum_{j=0}^{k} p_i(n)y(n+k-i, n_0, c(n+k-i)) \\ &\quad = \sum_{i=0}^{k} p_i(n) \sum_{j=1}^{k} c_j(n)y(n+k-i, n_0, E_j) + \sum_{j=1}^{k} \Delta c_j(n)y(n+k, n_0, E_j) \\ &\quad = g_n. \end{aligned}$$

Since $y(n, n_0, E_j)$ are solutions of (2.1.5), one has

$$\sum_{j=1}^{k} \Delta c_j(n)y(n+k, n_0, E_j) = g_n. \tag{2.2.7}$$

The equations (2.2.6) and (2.2.7) form a linear system of k equations in k unknowns $c_j(n)$, whose coefficient matrix is the Casorati matrix $K(n+1)$. The solution is given by

$$\begin{pmatrix} \Delta c_1(n) \\ \Delta c_2(n) \\ \vdots \\ \Delta c_k(n) \end{pmatrix} = K^{-1}(n+1) \begin{pmatrix} 0 \\ 0 \\ \vdots \\ g_n \end{pmatrix}. \tag{2.2.8}$$

Denoting by $M_{ik}(n+1)$ the (i, k) element of the adjoint matrix of $K(n+1)$, (2.2.8) becomes

$$\Delta c_i(n) = \frac{M_{ik}(n+1)}{\det K(n+1)} g_n, \qquad i = 1, 2, \ldots, k, \tag{2.2.9}$$

from which it follows that

$$c_i(n) = \Delta^{-1} \frac{M_{ik}(n+1)}{\det K(n+1)} g_n + \omega_i, \qquad c_i(n_0) = c_i.$$

By substituting the values of $c_i(n)$ in (2.2.3), we see that $z(n, n_0, c) = y(n, n_0, c(n))$ satisfies (2.1.4).

2.3. Linear Equations with Constant Coefficients

If in equation (2.1.1) the coefficients $p_i(n)$ are constants with respect to n, we obtain an important class of difference equations

$$\sum_{i=0}^{k} p_i y_{n+k-i} = g_n \qquad p_0 = 1. \tag{2.3.1}$$

The corresponding homogeneous equation is

$$\sum_{i=0}^{k} p_i y_{n+k-i} = 0. \tag{2.3.2}$$

Theorem 2.3.1. *The equation* (2.3.2) *has the solution of the form*

$$y_n = z^n, \tag{2.3.3}$$

where $z \in \mathbb{C}$, *and satisfies* (2.3.4).

Proof. Substituting (2.3.3) in (2.3.2), we have

$$z^n \sum_{i=0}^{k} p_i z^{k-i} = 0$$

from which it follows that

$$\sum_{i=0}^{k} p_i z^{k-i} = 0. \tag{2.3.4}$$

Equation (2.3.4) is a polynomial and it has k solutions in the complex field. Furthermore, it is said to be the characteristic equation of (2.3.2), and the polynomial $p(z) = \sum_{i=0}^{k} p_i z^{k-i}$ is called the characteristic polynomial. ■

Theorem 2.3.2. *If the roots $z_1, z_2, \ldots, z_k$ of $p(z)$ are distinct, then $z_1^n, z_2^n, \ldots, z_k^n$ are linearly independent solutions of* (2.3.2).

Proof. It is easy to verify that in this case the Casorati determinant is proportional to the determinant of the matrix

$$V(z_1, z_2, \ldots, z_k) = \begin{pmatrix} 1 & 1 & 1 \\ z_1 & z_2 & z_k \\ z_1^2 & z_2^2 & z_k^2 \\ \cdots & \cdots & \cdots \\ \cdots & \cdots & \cdots \\ z_1^{k-1} & z_2^{k-1} & z_k^{k-1} \end{pmatrix}, \tag{2.3.5}$$

which is known as Cauchy–Vandermonde matrix (or the Vandermonde matrix). Its determinant is given by

$$\det V(z_1, z_2, \ldots, z_k) = \prod_{i>j} (z_i - z_j), \tag{2.3.6}$$

which is different from zero if $z_i \neq z_j$ for all i and j.

From Theorem 2.1.4 it follows that if the roots of $p(z)$ are distinct, any solution of (2.3.2) can be expressed in the form

$$y_n = \sum_{i=1}^{k} c_i z_i^n. \tag{2.3.7}$$

When $p(z)$ has multiple roots, the solutions z_i^n corresponding to distinct roots are linearly independent. But they are not enough to form a base in S. It is possible however, to find other solutions and to form a base. ■

Theorem 2.3.3. *Let m_s be the multiplicity of the root z_s of $p(z)$. Then the functions*

$$y_s(n) = u_s(n) z_s^n, \tag{2.3.8}$$

where $u_s(n)$ are generic polynomials in n whose degree does not exceed $m_s - 1$, are solutions of (2.3.2) *and they are linearly independent.*

Proof. If z_s has multiplicity m_s as a root of $p(z)$, we have

$$p(z_s) = 0,\ p'(z_s) = 0, \ldots, \text{ and } p^{(m_s-1)}(z_s) = 0. \tag{2.3.9}$$

Let us look for a solution of (2.3.2) of the form

$$y_n = u_s(n) z_s^n. \tag{2.3.10}$$

By substituting, one gets

$$\sum_{i=0}^{k} p_i u_s(n+k-i) z_s^{k-i} = 0. \tag{2.3.11}$$

By the relation (1.1.4), we get

$$u_s(n+k-i) = \sum_{j=0}^{k-i} \binom{k-i}{j} \Delta^j u_s(n) \tag{2.3.12}$$

and, from (2.3.11), we have

$$\sum_{i=0}^{k} p_i z_s^{k-i} \sum_{j=0}^{k-i} \binom{k-i}{j} \Delta^j u_s(n) = \sum_{j=0}^{k} \Delta^j u_s(n) \sum_{i=0}^{k-j} \binom{k-i}{j} p_i z_s^{k-i}$$

$$= \sum_{j=0}^{k} \Delta^j u_s(n) z_s^j \frac{p^{(j)}(z_s)}{j!}. \tag{2.3.13}$$

In view of (2.3.9), one has it that the terms of (2.3.13) corresponding to $j = 0, 1, \ldots, m_s - 1$ are zero for all functions $u(n)$. To make the other $k - m_s + 1$ terms equal to zero, it is necessary that $\Delta^j u_s(n) = 0$ for $m_s \leq j \leq k$. This can be accomplished by taking $u_s(n)$ as a polynomial of degree not greater than $m_s - 1$. The proof that they are linearly independent is left as an exercise. ■

Corollary 2.3.1. *The general solution of* (2.3.2) *is given by*

$$y_n = \sum_{i=1}^{d} a_i u_i(n) z_i^n = \sum_{i=1}^{d} a_i \sum_{j=0}^{m_i-1} c_j n^j z_i^n = \sum_{i=1}^{d} \sum_{j=0}^{m_i-1} A_{ij} n^i z_i^n, \tag{2.3.14}$$

where $A_{ij} = \alpha_i c_j$, *and* d *is the number of distinct roots.*

The next theorem is useful in recognizing if a sequence y_n, $n \in \mathbf{N}_{n_0}^+$ is the solution of a difference equation.

Theorem 2.3.4. *A sequence* y_n, $n \in \mathbf{N}_{n_0}^+$ *satisfies the equation* 2.3.2 *iff, for all* $n \in \mathbf{N}_{n_0}^+$,

$$D(y_n, y_{n+1}, \ldots, y_{n+k}) = \det \begin{pmatrix} y_n & y_{n+1} & & y_{n+k} \\ y_{n+1} & y_{n+2} & \cdots & y_{n+k+1} \\ y_{n+k} & y_{n+k+1} & \cdots & y_{n+2k} \end{pmatrix} = 0$$

and moreover, $D(y_n, y_{n+1}, \ldots, y_{n+k-1}) \neq 0$.

Proof. Suppose that y_n satisfies 2.3.2. One has $y_{n+k+j} = -\sum_{i=1}^{k} p_i y_{n+k+j-i}$. By substituting in the last row of $D(y_n, y_{n+1}, \ldots, y_{n+k})$, one easily obtains that $D(y_n, \ldots, y_{n+k}) = 0$. Conversely, if this determinant is zero, one has, by developing with respect to the first row $\sum_{i=0}^{k} y_{n+k-i}A_i(n) = 0$, where $A_i(n)$ are the cofactors of the k-i^{th} elements. If in D one substitutes to the first row, the second one, and then the other rows, one obtains determinants identically zero. By developing these determinants again with respect to the elements of the first row one obtains:

$$\sum_{i=0}^{k} y_{n+j+k-i}A_i(n) = 0 \qquad j = 1, 2, \ldots, k-1.$$

The determinant $A_k(n)$ is not zero by hypothesis. One has, setting $p_i(n) = \dfrac{A_i(n)}{A_k(n)}$

$$y_n = -\sum_{i=0}^{k} p_i(n) y_{n+k-i},$$

$$y_{n+1} = -\sum_{i=0}^{k} p_i(n) y_{n+1+k-1},$$

$$y_{n+k-1} = -\sum_{i=0}^{n} p_i(n) y_{n+2k-1-i}.$$

By setting $n + 1$ instead of n in the first relation and subtracting the second one, we have

$$0 = -\sum_{i=0}^{k} \Delta p_i(n) y_{n+k+1-i}$$

and proceeding similarly for the others, one arrives at an homogeneous system of equations

$$\sum_{i=1}^{k} \Delta p_i(n) y_{n+k+j-i} = 0 \qquad j = 1, 2, \ldots, k-1,$$

whose determinant is not zero by hypothesis. It follows that the solution is $\Delta p_i(n) = 0$, that is the p_i are constant with respect to n and then the conclusion follows. ■

Example 1. Consider the equation

$$y_{n+1} - a y_n = 0, \qquad a \in \mathbb{C}. \tag{2.3.15}$$

The characteristic polynomial is $p(z) = z - a$ and its unique root is $z = a$. The general solution of (2.3.15) is then $y_n = ca^n$.

Example 2. Consider the equation

$$y_{n+2} - y_{n+1} - y_n = 0. \tag{2.3.16}$$

The characteristic polynomial is $p(z) = z^2 - z - 1$, which has roots $z_1 - \dfrac{1+\sqrt{5}}{v}$; $z_2 = \dfrac{1-\sqrt{5}}{2}$.

Therefore, the general solution of (2.3.16) is

$$y_n = c_1\left(\frac{1+\sqrt{5}}{2}\right)^n + c_2\left(\frac{1-\sqrt{5}}{2}\right)^n,$$

and is known as the Fibonacci sequence.

Example 3. Consider the equation

$$y_{n+1} - ay_n = g_n.$$

From Example 1, it follows that the general solution of the homogeneous equation is ca^n. Applying the method of variation of constants it follows that

$$(c(n+1) - c(n))a^{n+1} = g_n$$

and

$$c(n) = \Delta^{-1}\frac{g(n)}{a^{n+1}} = \sum_{j=0}^{n-1}\frac{g(j)}{a^{j+1}},$$

from which, we obtain

$$y_n = y_0 a^n + \sum_{j=0}^{n-1} g(j)a^{n-j-1}.$$

Example 4. The following equation often arises in discretization of second order differential equations, namely,

$$y_{n+2} - 2qy_{n+1} + y_n = f_n, \tag{2.3.17}$$

where $q \in \boldsymbol{C}$. The homogeneous equation has the general solution $y_n = c_1 z_1^n + c_2 z_2^n$, where z_1 and z_2 are the distinct roots of the second degree equation $z^2 - 2qz + 1 = 0$. It is useful, in the applications, to write the general solution in two different forms. In the first form, the linearly independent solutions

$$y_n^{(1)} = \frac{z_2 z_1^n - z_1 z_2^n}{z_2 - z_1}; \qquad y_n^{(2)} = \frac{z_2^n - z_1^n}{z_2 - z_1}$$

are used, which give, as a general solution of the homogeneous equation

$$y_n = c_1 y_n^{(1)} + c_2 y_n^{(2)}. \tag{2.3.18}$$

In the second case, one uses the Chebyshev polynomials (see Appendix C) $T_n(q)$ and $U_n(q)$ as linearly independent solutions, obtaining

$$y_n = c_1 T_n(q) + c_2 U_{n-1}(q). \tag{2.3.19}$$

The advantage of using (2.3.18) is that the base $y_n^{(1)}$, $y_n^{(2)}$ is a canonical one, that is,

$$y_0^{(1)} = 1, \qquad y_0^{(2)} = 0,$$
$$y_1^{(1)} = 0, \qquad y_1^{(2)} = 1,$$

from which it follows that for the initial value problem (2.3.18) wc have

$$y_n = y_0 y_n^{(1)} + y_1 y_n^{(2)}.$$

The advantage of the form (2.3.19) lies in the fact that the functions $T_n(q)$ and $U_n(q)$ have many interesting properties that make their use especially helpful in Numerical Analysis and Approximation Theory.

The solution of (2.3.17) can then be written in the following form,

$$y_n = y_0 y_n^{(1)} + y_1 y_n^{(2)} + \sum_{j=0}^{n-2} y_{n-j-1}^{(2)} \cdot f_j.$$

Example 5. Consider the equation

$$p_0 y_{n+2} + p_1 y_{n+1} + p_2 y_n = 0. \tag{2.3.20}$$

The solution can be written in terms of the roots of the polynomial $p_0 z^2 + p_1 z + p_2 = 0$ as usual.

It is interesting, however, to give the solution in terms of Chebyshev polynomials. Suppose $p_0 p_2 > 0$ and let $\rho = \left(\frac{p_2}{p_0}\right)^{1/2}$ and $q = -\frac{p_1}{2(p_0 p_2)^{1/2}}$. One easily verifies that $\rho^n T_n(q)$ and $\rho^n U_n(q)$ are solutions of (2.3.20). It follows then that $y_n = c_1 \rho^n T_n(q) + c_2 \rho^n U_{n-1}(q)$ is the general solution.

2.4. *Use of Operators Δ and E*

The method of solving difference equations with constant coefficients becomes simple and elegant when we use the operators Δ and E.

Using the operator E equations (2.3.1) and (2.3.2) can be rewritten in the form

$$p(E) y_n = g_n, \tag{2.4.1}$$

$$p(E) y_n = 0, \tag{2.4.2}$$

where

$$p(E) = \sum_{i=0}^{k} p_i E^{k-i}. \tag{2.4.3}$$

It is immediate to verify that

$$p(E)z^n = z^n p(z) \quad \text{and} \quad p(E) = \prod_{i=1}^{k} (E - z_i I), \tag{2.4.4}$$

where $z_1, z_2, \ldots, z_k$ are the zeros of $p(z)$.

If there are s distinct roots with multiplicity m_j, $j = 1, 2, \ldots, s$, then $p(E)$ can be written as $p(E) = \prod_{i=1}^{s} (E - z_i I)^{m_i}$ and (2.4.2) becomes

$$\prod_{i=1}^{s} (E - z_i I)^{m_i} y_n = 0, \tag{2.4.5}$$

from which it is seen that the homogeneous equation can be split into s difference equations of order m_j. In fact, the commutability of the operators $(E - z_i I)$ implies the following result.

Proposition 2.4.1. *The solution x_n of the equation*

$$(E - z_j I)^{m_j} x_n = 0 \tag{2.4.6}$$

is a solution of (2.4.5). ■

The problem simplifies further since it is possible to define the inverses of $(E - z_i I)$, $i = 1, 2, \ldots, k$.

Proposition 2.4.2. *Let $\{y_n\}$ be a sequence and let $f(z) = \sum_{i=0}^{m} a_i z^i$ be a polynomial of degree m. Then,*

$$f(E)(z^n y_n) = z^n f(zE) y_n. \tag{2.4.7}$$

Proof. By definition of $f(z)$, it follows that

$$f(E)(z^n y_n) = \sum_{i=0}^{m} a_i E^i (z^n y_n) = \sum_{i=0}^{m} a_i z^{n+i} E^i y_n$$

$$= z^n \sum_{i=0}^{m} a_i (z^i E^i) y_n = z^n f(zE) y_n. \quad ■$$

Definition 2.4.1. The inverse of the operator $(E - zI)$ is the operator $(E - zI)^{-1}$ such that

$$(E - zI)(E - zI)^{-1} = I.$$

Theorem 2.4.1. *Let $z \in \mathbb{C}$. Then, the inverse of $E - zI$ is given by*

$$(E - zI)^{-1} = z^{n-1}\Delta^{-1}z^{-n}. \tag{2.4.8}$$

Proof. Applying $(E - zI)$ to both sides of (2.4.8) and using the result of Proposition 2.4.2, one gets

$$\begin{aligned}(E - zI)(E - zI)^{-1} &= (E - zI)z^{n-1}\Delta^{-1}z^{-n} \\ &= z^{n-1}z(E - I)\Delta^{-1}z^{-n} \\ &= z^n \cdot z^{-n}I = I. \qquad \blacksquare\end{aligned}$$

Corollary 2.4.1. *For $m = 1, 2, \ldots,$ one has*

$$(E - zI)^{-m} = z^{n-m}\Delta^{-m}z^{-n}. \tag{2.4.9}$$

The equation (2.4.9) allows us to find very easily the solutions of (2.4.6) and then of (2.4.1). In fact, from (2.4.6) and (2.4.9) we have

$$x_n = (E - z_j)^{-m_j} \cdot 0 = z_j^{n-m_j}\Delta^{-m_j} \cdot 0.$$

But, we know that $\Delta^{-m_j} \cdot 0 = q_j(n)$, where $q_j(n)$ is a polynomial of degree less than m_j.

Hence, the solution x_n is given by $x_n = z_j^{n-m_j}q_j(n)$, and this can be repeated for $j = 1, 2, \ldots, s$.

Usually, because m_j is independent of n, one prefers to consider the previous solution multiplied by $z_j^{m_j}$. The solutions corresponding to the multiple root z_j are then $z_j^n q_j(n)$, and hence, the general solution of (2.4.1) is $y_n = \sum_{j=1}^{s} a_j q_j(n) z_j^n$, which is equivalent to (2.3.14).

In general, to get a solution of the nonhomogeneous equation (2.4.1), one can proceed as described in the previous section by applying the method of variation of constants. Usually this way of proceeding is too long and in some cases can be avoided using the definition of $p^{-1}(E)$.

Proposition 2.4.3. *Let $f(z)$ be a polynomial of degree k and $z \in \mathbb{C}$ with $f(z) \neq 0$. Then*

$$f^{-1}(E)z^n = \frac{z^n}{f(z)}. \tag{2.4.10}$$

Proof. By applying $f(E)$ to both sides of (2.4.10), one obtains

$$f(E)f^{-1}(E)z^n = f(E)\frac{z^n}{f(z)} = z^n. \qquad \blacksquare$$

Proposition 2.4.4. *Let $f(z)$ be a polynomial of degree k and $z_1 \in \mathbb{C}$ be a root of multiplicity m. Then, setting $g(z) = (z - z_1)^{-m} f(z)$, one has*

$$f^{-1}(E) z_1^n = \frac{z_1^{n-m} n^{(m)}}{g(z_1) m!} \tag{2.4.11}$$

Proof. By applying $f(E)$ to both sides and using (2.4.7), one obtains

$$f(E) f(E^{-1}) z_1^n = \frac{z_1^{n-m} f(z_1 E) n^{(m)}}{g(z_1) m!} = z_1^n \frac{g(z_1 E)(E - I)^m n^{(m)}}{g(z_1) m!}$$
$$= \frac{z_1^n g(z_1 E) \cdot 1}{g(z_1)} = z_1^n. \quad \blacksquare$$

Proposition 2.4.5. *Let $f(z)$ be a polynomial of degree k and y_n be a sequence. Then for every $n \in \mathbf{N}$, we have $f^{-1}(E) z^n y_n = z^n f^{-1}(zE) y_n$.* $\blacksquare$

These results can be used to obtain particular solutions of the equation $p(E) y_n = g(n)$.

Let us consider the most frequent cases:

(a) $g(n) = g$ constant. If $p(1) \neq 0$ from (2.4.10) one obtains

$$\bar{y}_n = p^{-1}(E) g = g p^{-1}(E) \cdot 1^n = \frac{g}{p(1)} = \frac{g}{\sum_{i=0}^{k} p_i}$$

(b) $g(n) = \sum_{i=1}^{s} a_i z_i^n$ with $p(z_i) \neq 0$. From (2.4.10) one has

$$\bar{y}_n = p^{-1}(E) \sum_{i=1}^{s} a_i z_i^n = \sum_{i=1}^{s} a_i p^{-1}(E) z_i^n = \sum_{i=1}^{n} a_i \frac{z_i^n}{p(z_i)}.$$

(c) Same as in (b), but z_j is a root of $p(z)$ of multiplicity m. From (2.4.10) and (2.4.11) we obtain

$$\bar{y}_n = \sum_{i \neq j} \frac{a_i z_i^n}{f(z_i)} + a_j \frac{z_j^{n-m} n^{(m)}}{g(z_j) m!},$$

where $g(z) = \dfrac{f(z)}{(z - z_j)^m}$.

(d) $g(n) = e^{in\alpha}$. This case can be treated as in (b) or (c) by putting $z = e^{i\alpha}$.

(e) $g(n) = \cos n\alpha$, $g(n) = \sin n\alpha$. We can proceed as in case (d) taking the real or imaginary part.

2.5. *Method of Generating Functions*

The method of generating functions is another elegant method for solving linear difference equations with constant coefficients. Its importance is growing in discrete mathematics.

Definition 2.5.1. Given a sequence $\{y_n\}$, we shall call a formal series generated by it the expression

$$Y = \sum_{i=0}^{\infty} y_i x^i, \tag{2.5.1}$$

where x is a symbol.

Only in the case where x will be a complex value, the problem of convergence of (2.5.1) will arise. The formal series are often called generating functions of $\{y_n\}$. In the set of all formal series, we can define operations that make such a set algebraically similar to the set of rational numbers.

Definition 2.5.2. Given two formal series Y and Z, their sum is defined by

$$Y + Z = \sum_{n=0}^{\infty} (y_n + z_n)x^n.$$

Definition 2.5.3. The product of two formal series Y and Z is given by

$$YZ = \sum_{n=0}^{\infty} c_n x^n, \tag{2.5.2}$$

where

$$c_n = \sum_{i=0}^{n} y_i z_{n-i} = \sum_{i=0}^{n} z_i y_{n-i}. \tag{2.5.3}$$

We list some simple properties of formal series:

(i) The product of two formal series is commutative;
(ii) Given three formal series Y, Z, T we have $(Y+Z)T = YT + ZT$;
(iii) The unit element with respect to the product is the formal series $I = 1 + 0x + 0x^2 + \dots$;
(iv) The zero element is the formal series $0 + 0x + 0x^2 + \dots$;
(v) The set of all the formal series is an integral domain;
(vi) Let $Y = \sum_{i=0}^{\infty} y_i x^i$. If $y_0 \neq 0$ then there exists the formal series Y^{-1} such that $Y^{-1}Y = I$.

The polynomials are particular formal series with a finite number of terms.

Consider now the linear difference equation

$$\sum_{i=0}^{k} p_i y_{n+k-i} = 0, \qquad \text{with } p_0 = 1, \tag{2.5.4}$$

and we shall associate with it the two formal series

$$P = p_0 + p_1 x + \dots + p_k x^k, \text{ and}$$
$$Y = y_0 + y_1 x + \dots. \tag{2.5.5}$$

P is different from the characteristic polynomial. In fact, one has

$$P = x^k p\left(\frac{1}{x}\right) \equiv \bar{p}(x), \tag{2.5.6}$$

where $p(z)$ is the characteristic polynomial. The product Q of the two series is

$$Q = YP = q_0 + q_1 x + \ldots + q_{k-1}x^{k-1} + q_k x^k + \ldots \tag{2.5.7}$$

where

$$q_n = \sum_{i=0}^{n} p_i y_{n-i}. \tag{2.5.8}$$

In view of (2.5.4) it is easy to see that $0 = q_k = q_{k+1} = \ldots$, which means that Q is a formal series with a finite number of terms. Moreover, because P is invertible ($p_0 = 1$), one has

$$Y = P^{-1}Q. \tag{2.5.9}$$

If we consider the symbol x as an element z of the complex plane, then (2.5.9) gives the values of Y as ratio of two polynomials

$$Y = \frac{q(z)}{\bar{p}(z)} = \frac{q(z)}{z^k p\left(\frac{1}{z}\right)}, \tag{2.5.10}$$

where $p\left(\frac{1}{z}\right)$ is the characteristic polynomial and $q(z) = \sum_{i=0}^{k-1} q_i z^i$. The roots of $\bar{p}(z)$ are z_i^{-1}, where z_i are the roots of $p(z)$.

It is known in the theory of complex variables that every expression like (2.5.10) is equal to the sum of the principal parts of its poles. The poles of (2.5.10) are the roots $\frac{1}{z_i}$ of the denominators, and therefore,

$$\begin{aligned} Y \equiv \sum_{n=0}^{\infty} y_n z^n &= \sum_{i=1}^{s} \sum_{j=1}^{m_i} a_{ij}(z - z_i^{-1})^{-j} \\ &= \sum_{i=1}^{s} \sum_{j=1}^{m_i} a_{ij}(-1)^j z_i^j (1 - z_i z)^{-j}, \end{aligned} \tag{2.5.11}$$

where s is the number of distinct roots of $p(z)$ and m_j their multiplicity. The coefficients a_{ij} are the coefficients in the Laurent series of (2.5.10). For $|z_i z| < 1$, $i = 1, 2, \ldots, k$, $(1 - z_i z)^{-j}$ can be expressed as

$$(1 - z_i z)^{-j} = \left(\sum_{n=0}^{\infty} z_i^n z^n\right)^j = \sum_{n=0}^{\infty} \binom{n+j-1}{n} z_i^n z^n, \tag{2.5.12}$$

and substituting in (2.5.11), we get

$$\sum_{n=0}^{\infty} y_n z^n = \sum_{n=0}^{\infty} z^n \sum_{i=1}^{s} \sum_{j=1}^{m_i} a_{ij}(-1)^{-j} \binom{n+j-1}{n} z_i^{n+j}.$$

If we write

$$q_i(n) = \sum_{j=1}^{m_i} a_{ij} \binom{n+j-1}{n} (-1)^j z_i^{-j},$$

we obtain

$$y_n = \sum_{i=1}^{s} q_i(n) z_i^n, \tag{2.5.13}$$

which is equivalent to (2.3.14).

Theorem 2.5.1. *Suppose that the roots of the characteristic polynomial $p(z)$ are inside the unit disk of the complex plane. Then the formal series Y converges inside the unit disk.*

Proof. The polynomial $\bar{p}(z)$ has zeros outside the unit disk and $q(z)/\bar{p}(z)$ has no poles in it. Y must then coincide with the Taylor series in the unit disk. ∎

Theorem 2.5.2. *If the characteristic polynomial $p(z)$ has no roots outside the unit disk and those on the unit circle are simple, then the coefficients y_n of Y are bounded.*

Proof. From (2.5.13), it follows that $q_i(n)$ corresponding to $|z_i| = 1$ are constants with respect to n.

The method of generating functions can also be used to obtain solutions of nonhomogeneous equation (2.1.4). One can proceed as before with the difference that $q_k, q_{k+1}, \ldots,$ are not zero.

In fact from (2.5.8) and (2.1.4) one has $q_k = \sum_{i=0}^{k} p_i y_{k+i} = g_0$ and, in general, for $n = 1, 2, \ldots$

$$q_{n+k} = \sum_{i=0}^{k} p_i y_{n+k-i} = g_n, \tag{2.5.14}$$

since $p_i = 0$ for $i > k$. The series (2.5.7) can be written as

$$\begin{aligned} Q &= \sum_{n=0}^{k-1} z^n \sum_{i=0}^{n} p_i y_{n-i} + z^k \sum_{n=0}^{\infty} g_n z^n \\ &= Q_1(z) + z^k Q_2(z), \end{aligned} \tag{2.5.15}$$

where as (2.5.9) becomes

$$\sum_{i=0}^{\infty} y_i z^i = \frac{Q_1(z) + z^k Q_2(z)}{\bar{p}(z)}. \quad \blacksquare \tag{2.5.16}$$

The polynomial $Q_1(z)$ depends only on the initial values $y_0, y_1, \ldots, y_{k-1}$, while $Q_2(z)$ is a formal series defined by the sequence $\{g_n\}$. Proceeding as in (2.5.10), (2.5.11) and (2.5.12) one obtains the solution $\{y_n\}$.

This procedure can be further simplified by considering that inside the region of convergence, it represents a function $f(z)$, which is said to be the transformed function of the sequence $\{y_n\}$. For example, in the unit disk, the function $(1-z)^{-1}$ is the transformed function of the constant sequence $\{1\}$, since

$$(1-z)^{-1} = \sum_{i=0}^{\infty} z^i.$$

On Table 1, transformed functions of some important sequences are given.

Now suppose that $Q_2(z)$ is the transformed function of $\{g_n\}$. After doing the necessary algebraic operations, one obtains from (2.5.16), $\sum_{i=0}^{\infty} y_i z^i = G(z)$, where $G(z)$ is the function resulting in the right-hand side of (2.5.16).

By expanding $G(z)$ in Taylor series and equating the coefficients of the powers of the same orders on both sides, we arrive at the solution $\{y_n\}$.

Example. Consider the equation $y_{n+1} + y_n = -(n+1)$, $y_0 = 1$. Here $Q_1(z) = 1$, $Q_2(z) = -\sum_{n=0}^{\infty} (n+1)z^n = -\dfrac{1}{(1-z)^2}$ and

$$G(z) = \frac{1 - \dfrac{z}{(1-z)^2}}{1+z} = \frac{1-3z+z^2}{(1+z)(1-z)^2} = \frac{5}{4}\frac{1}{1+z} - \frac{1}{4}\frac{1}{1-z} - \frac{1}{2}\frac{z}{(1-z)^2}.$$

From Table 2.1, we find that $\dfrac{1}{1+z} = \sum_{n=0}^{\infty} (-1)^n z^n$, $\dfrac{1}{1-z} = \sum_{n=0}^{\infty} z^n$, and $\dfrac{z}{(1-z)^2} = \sum_{n=0}^{\infty} nz^n$. Therefore, $\sum_{n=0}^{\infty} y_n z^n = \sum_{n=0}^{\infty} \left[\frac{5}{4}(-1)n - \frac{1}{4} - \frac{1}{2}n\right] z^n$, from which we obtain $y_n = \dfrac{5}{4}(-1)^n - \dfrac{1}{4} - \dfrac{1}{2}n$.

In some applications, especially in system theory, instead of generating functions defined in (2.5.1), generating functions called Z transform defined by $X(z) = \sum_{n=0}^{\infty} y_n z^{-n}$ are used. It is evident that $X(z) = Y\left(\dfrac{1}{z}\right)$ and therefore,

$$X(z) = \frac{Q_1\left(\frac{1}{z}\right) + z^{-k} Q_2\left(\frac{1}{z}\right)}{\bar{p}\left(\frac{1}{z}\right)} = \frac{Q_1\left(\frac{1}{z}\right) + z^{-k} Q_2\left(\frac{1}{z}\right)}{z^{-k} p(z)} = \frac{z^k Q_1\left(\frac{1}{z}\right) + Q_2\left(\frac{1}{z}\right)}{p(z)},$$

TABLE 1.

y_n	$f(z)$	Domain of Convergence
1	$(1-z)^{-1}$	$\|z\|<1$
n	$z(1-z)^{-2}$	
$(n+m)^{(m)}$	$m!(1-z)^{-m-1}$	
$n^{(m)}$	$m!z^m(1-z)^{-m-1}$	
n^m	$zp_m(z)(1-z)^{-n-1}(*)$	
k^n	$(1-kz)^{-1}$	$\|z\|<k^{-1}$
$(n+m)^{(m)}k^n$	$m!(1-kz)^{-m-1}$	
e^{an}	$(1-e^a z)^{-1}$	$\|z\|<e^{-a}$
$k^n \cos an$	$\dfrac{1-kz\cos a}{1-2kz\cos a+k^2z^2}$	$\|z\|<k^{-1}$
$k^n \sin an$	$\dfrac{kz\sin a}{1-2kz\cos a+k^2z^2}$	$\|z\|<k^{-1}$
$\dfrac{B_n}{n!}$	$\dfrac{z}{e^z-1}$	$\|z\|<2\pi$
$\binom{n}{m}$	$z^m(1-z)^{-m-1}$	$\|z\|<1$
$\binom{k}{n}$	$(1+z)^k$	$\|z\|<1$
$(-1)\cdot\binom{k}{n}$	$(1-z)^k$	$\|z\|<1$
$U_n(x)$ (**)	$(1-2xz+z^2)^{-1}$	$\|x\|\le 1, \|z\|<1$
$T_n(x)$	$(1-xz)(1-2xz+z^2)^{-1}$	

(*) $p_m(z)$ is a polynomial of degree m satisfying the recurrence relation $p_{m+1}(z) = (mz+1)p_m(z) + z(1-z)p'_m(z)$, $p_1 = 1$.
(**) $T_n(x)$ and $U_n(x)$ are the Chebyshev polynomials (see Appendix C).

where $Q_2\left(\frac{1}{z}\right)$ is the Z transform of $\{g_n\}$ and $p(z)$ is the characteristic polynomial. Using the table of Z transforms, everything goes similarly as before.

2.6. Stability of Solutions

The stability problem will be studied in a more general setting in a later chapter. In this section we shall consider only the stability problem for linear difference equations, which is very important in applications.

Definition 2.6.1. The solution $\bar{y}_n$ of (2.1.4) is said to be stable if, for any other solution y_n of (2.1.4), the following difference is bounded:

$$l_n = y_n - \bar{y}_n, \qquad n \in \mathbf{N}^+_{n_0}. \tag{2.6.1}$$

Definition 2.6.2. The solution $\bar{y}_n$ of (2.1.4) is said to be asymptotically stable if for any other solution y_n of (2.1.4), one has $\lim_{n\to\infty} l_n = 0$.

Definition 2.6.3. The solution $\bar{y}_n$ of (2.1.4) is said to be unstable if it is not stable.

From Lemma 2.1.3 it follows that the difference l_n satisfies the homogeneous equations (2.1.5). In the case of linear equations with constant coefficients we have the following results.

Therorem 2.6.1. *The solution y_n of* (2.3.1) *is asymptotically stable if the roots of the characteristic polynomial are within the unit circle in the complex plane.*

Proof. From (2.3.14) we have

$$\lim_{n\to\infty} |y_n - \bar{y}_n| = \lim_{n\to\infty} \sum_{i=1}^{s} \sum_{j=0}^{m_i-1} |A_{ij}|n^j|z_i^n|.$$

If $|z_i| < 1$ one has $\lim_{n\to\infty} |y_n - \bar{y}_n| = 0$ and vice versa. ■

Theorem 2.6.2. *The solution $\bar{y}_n$ of* (2.3.1) *is stable if the module of the roots of the characteristic polynomial is less than or equal to* 1 *and that those with modules equal to* 1 *are simple roots.*

Proof. From (2.3.14), it is evident that the terms coming from roots with modules less than 1 gives a vanishing contribution for $n \to \infty$, while the terms coming from roots with unit modules, give a bounded contribution to l_n, since $j = 0$. ■

It can happen that for some initial conditions, the solution remains bounded even in presence of multiple roots on the unit circle, as shown in the next example.

Example 1. Consider the equation

$$y_{n+2} - 2y_{n+1} + y_n = 0, \qquad y_0 = y_1 = c.$$

This equation admits the solution $y_n = c$.

Often in applications it becomes necessary to study the stability of a constant solution which exists, as we have seen, if g_n is constant.

Example 2. Consider the equation

$$y_{n+2} - y_{n+1} + \tfrac{1}{4}y_n = 2.$$

We have $p(z) = (z - \frac{1}{2})^2$ and all solutions will be asymptotically stable. In particular the constant solution $\bar{y} = 8$ is asymptotically stable.

In fact, the general solution is given by $y_n = (c_1 + c_2 n)2^{-n} + 8$ and $\lim_{n\to\infty} (y_n - 8) = 0$.

From definitions 2.6.1, 2.6.2 and 2.6.3, we see that the properties of stability and instability are usually referred with respect to a particular solution $\bar{y}_n$. In the case where all solutions tend to a unique solution $\bar{y}_n$ as $n \to \infty$, it is often said (especially in numerical analysis) that the difference equation itself (or the numerical method represented by it) is asymptotically stable. Moreover, in some branches of applications, a special terminology is used that is becoming more and more popular and it is worthwhile to mention it.

Definition 2.6.4. A polynomial with roots within the unit disk in the complex plane is called a Schur polynomial.

Definition 2.6.5. A polynomial with roots in the unit disk in the complex plane with only simple roots on the boundary is called a Von Neumann polynomial.

Using this terminology, the Theorems 2.6.1 and 2.6.2 can be restated as follows.

Theorem 2.6.1. *The solution $\bar{y}_n$ is asymptotically stable, if the characteristic polynomial is a Schur polynomial.*

Theorem 2.6.2. *The solution $\bar{y}_n$ is stable if the characteristic polynomial is a Von Neumann polynomial.*

2.7. Absolute Stability

One of the main application of linear difference equations is the study of discretization methods for differential equations. The difference equations, as we have seen, can be solved recursively. This is not possible for the differential equations and these are usually solved approximately using difference equations that satisfy some suitable conditions.

Let us consider

$$y' = f(t, y), \qquad y(t_0) = y_0, \tag{2.7.1}$$

where $t \in [t_0, T)$ and suppose that this continuous problem has a unique solution $y(t)$. Let $h > 0$ and $t_i = t_0 + ih$ with $i = 0, 1, \dots, N = \frac{T}{h}$.

Let the discrete problem approximating (2.7.1) be denoted by

$$F_h(y_n, y_{n+1}, \dots, y_{n+k}, f_n, f_n, \dots, f_{n+k}) = 0, \tag{2.7.2}$$

where $y_i = y(t_i) + 0(h^q)$, $q > 1$ and $i = 0, 1, \dots, k-1$, $n + k \leq N$. We suppose that (2.7.2) has a unique solution y_n. As the discrete problem is represented by a difference equation of order k, it needs k initial conditions, only one of which is given from the continuous problem. The others are approximately found in some way.

Definition 2.7.1. The problem (2.7.2) is said to be consistent with the problem (2.7.1) if

$$\begin{aligned} F_h(y(t_n), y(t_{n+1}), \dots, y(t_{n+k}), f(t_n, y(t_n)), \dots, f(t_{n+k}, y(t_{n+k}))) \\ \equiv \tau_n = 0(h^{p+1}) \end{aligned} \tag{2.7.3}$$

with $p \geq 1$.

The quantity τ_n is called the truncation error. The equation (2.7.3) can be considered as a perturbation of (2.7.2).

Definition 2.7.2. The discrete problem (2.7.2) is said to be convergent to the problem (2.7.1) if the solution y_n of (2.7.2) tends to the solution $y(t)$ of (2.7.1) for $n \to \infty$, and $t_n - t_0 = nh \leq T$.

Since the solution of the continuous problem satisfies (2.7.3), which is a perturbation of (2.7.2), the convergence will occur when (2.7.2) will be insensitive to such a perturbation, that is, when (2.7.2) is stable under perturbation. As a consequence, the consistency is not enough to guarantee the convergence.

We shall study the problem in some detail for the main class of methods called linear multistep methods (LMF). These methods are obtained when F_h is linear in its arguments, namely,

$$\sum_{i=0}^{k} \alpha_i y_{n+i} - h \sum_{i=0}^{k} \beta_i f_{n+i} = 0, \tag{2.7.4}$$

with $\alpha_k = 1$ and coefficients α_i, β_i are real numbers. Using the shift operator E and the two polynomials ρ and σ given by

$$\rho(z) = \sum_{i=0}^{k} \alpha_i z^i, \tag{2.7.5}$$

$$\sigma(z) = \sum_{i=0}^{k} \beta_i z^i, \tag{2.7.6}$$

equation (2.7.4) can be written as

$$\rho(E)y_n - h\sigma(E)f_n = 0. \tag{2.7.7}$$

The two polynomials $\rho(z)$ and $\sigma(z)$ characterize the method (2.7.2) uniquely and one often refers to them as (ρ, σ) method.

The relation (2.7.3) becomes

$$\rho(E)y(t_n) - h\sigma(E)f(t_n, y(t_n)) = \tau_n. \tag{2.7.8}$$

Theorem 2.7.1. *Suppose that f is smooth enough. Then the quantity τ_n is infinitesimal of order 2 with respect to h if the following two conditions are verified:*

$$\rho(q) \equiv \sum_{i=0}^{k} \alpha_i = 0, \tag{2.7.9}$$

and

$$\rho'(1) - \sigma(1) \equiv \sum_{i=0}^{k} i\alpha_i - \sum_{i=0}^{k} \beta_i = 0. \tag{2.7.10}$$

(For the proof of Theorem 2.7.1, see problem 2.18).

The conditions (2.7.9) and (2.7.10) are said to be consistency conditions.

If f is nonlinear, then the study of stability of (2.7.7) is in general difficult. Usually, one studies the behavior of solutions of (2.7.7) for particular linear functions f, which are called test functions. The most used test functions are

$$f(y) = 0, \tag{2.7.11}$$

and

$$f(y) = \lambda y, \qquad \operatorname{Re} \lambda \leq 0. \tag{2.7.12}$$

The use of test function (2.7.11) is justified by considering that in (2.7.7) the values of f are multiplied by h and then, in the limit as $h \to 0$, the contribution to solutions of the terms containing f_{n+i} can be disregarded. Also, one sees that the methods give good results when applied to the simple equation $y' = 0$.

The use of test function (2.7.12) is justified by considering that in the neighborhood of an asymptotically stable solution of (2.7.7), the first order approximation theorem says that the behavior of any solution is established by the linear part that looks like (2.7.12).

Let us first consider the test equation (2.7.11). Then (2.7.7) becomes

$$\rho(E)y_n = 0. \tag{2.7.13}$$

Definition 2.7.3. The method (ρ, σ) is said to be 0-stable if the solution $y_n = 0$, $n \in \boldsymbol{N}^+$ of (2.7.13) is stable.

As a simple consequence of Theorem 2.6.1, we have the following result.

Theorem 2.7.2. *The method* (ρ, σ) *is* 0*-stable if* $\rho(z)$ *is a Von-Neumann polynomial.*

Theorem 2.7.3. *The method* (ρ, σ) *is convergent in the finite interval* $(0, T)$ *iff it is consistent and* 0*-stable.*

Proof. Let us write $f(t_n, y(t_n)) - f_n = \boldsymbol{C}_n e_n$, where $e_n = y(t_n) - y_n$. Then, subtracting (2.7.4) from (2.7.8), one obtains the error equation

$$\rho(E)e_n = h\sigma(E)\boldsymbol{C}_n e_n + \tau_n \equiv g_n,$$

with

$$e_j = O(h^q), \qquad q \geq 1, \qquad j = 0, 1, \ldots, q-1.$$

The necessity part of the proof will be left as an exercise (see problem 2.20 and 2.21). Suppose now that the method is 0-stable and consistent, we shall prove the convergence. We will use the formal series method, from which we get

$$\sum_{n=0}^{\infty} e_n z^n = \frac{Q_1(z) + z^k Q_2(z)}{\bar{p}(z)} = [Q_1(z) + z^k Q_2(z)]Q_3(z),$$

where

$$Q_1(z) = \sum_{i=0}^{k-1} q_i^{(1)} z^i, \qquad q_i^{(1)} = \sum_{j=0}^{i} \alpha_j e_{i-j},$$

$$Q_2(z) = \sum_{i=0}^{\infty} q_i^{(2)} z^i, \qquad q_i^{(2)} = g_i$$

$$Q_3(z) = \sum_{n=0}^{\infty} \gamma_n z^n.$$

By Theorem 2.5.2 and by 0-stability we see that γ_n are bounded. By multiplying the formal series we get

$$Q_1(z)Q_3(z) = \sum_{n=0}^{\infty} \delta_n^{(1)} z^n, \qquad \delta_n^{(1)} = \sum_{i=0}^{s} q_i^{(1)} \gamma_{n-i}, \qquad s = \min(n, k-1),$$

$$z^k Q_w(z)Q_3(z) = \sum_{n=0}^{\infty} \delta_n^{(2)} z^n, \qquad \delta_n^{(2)} = \begin{cases} 0, \; n = 0, 1, \ldots, k-1 \\ \sum_{i=0}^{n-k} q_i^{(2)} \gamma_{n-k-i}, & n \geq k. \end{cases}$$

By equating the coefficients, we have $e_{n+k} = \delta_{n+k}^{(1)} + \delta_{n+k}^{(2)}$ and $|e_{n+k}| \leq |\delta_{n+k}^{(1)}| + |\delta_{n+k}^{(2)}|$. But

$$|\delta_{n+k}^{(1)}| \leq \Gamma \sum_{i=0}^{k-1} |q_i^{(1)}| \leq \Gamma \sum_{i=0}^{k-1} \Big|\sum_{j=0}^{i} \alpha_j e_{i-j}\Big| \leq \Gamma H \sum_{i=0}^{k-1} \sum_{j}^{i} |\alpha_j| = \Gamma H A$$

where

$$\Gamma = \max|\gamma_i|, \qquad H = \max_{0 \leq i \leq k-1} |e_i|, \qquad A = \sum_{i=0}^{k-1} \sum_{j=0}^{i} |\alpha_j| = \sum_{i=0}^{k-1} (k-i)|\alpha_i|,$$

and

$$\begin{aligned} |\delta_{n+k}^{(2)}| &\leq \Gamma \sum_{i=0}^{n} |q_i^{(2)}| \leq \Gamma\left(\sum_{i=0}^{n} |\tau_i| + h \sum_{i=0}^{n} |\sigma(E) C_i e_i| \right) \\ &\leq \Gamma\left(\theta_n + h \sum_{i=0}^{n} \sum_{j=0}^{k} |\beta_j||C_{i+j}||e_{i+j}| \right) \\ &\leq \Gamma\left(\theta_n + hL \sum_{i=0}^{n} \sum_{j=0}^{k} |\beta_j||e_{i+j}| \right), \end{aligned}$$

with $\theta_n = \sum_{i=0}^{n} |\tau_i|$ and $L = \max\limits_{\substack{0<i<n \\ 0 \leq j \leq k}} |C_{i+j}|$. Moreover,

$$\begin{aligned} |\delta_{n+k}^{(2)}| &\leq \Gamma\left(\theta_n + hL \sum_{s=0}^{n+k} |e_s| \sum_{r=r_1}^{r_2} |\beta_r| \right) \\ &\leq \Gamma\left(\theta_n + hL \sum_{s=0}^{n+k-1} |e_s| \sum_{r=r_1}^{r_2} |\beta_r| + hL|\beta_k||e_{n+k}| \right) \\ &\leq \Gamma\left(\theta_n + hLB \sum_{s=0}^{n+k-1} |e_s| + hL|\beta_k||e_{n+k}| \right), \end{aligned}$$

where

$$r = \begin{cases} 0, & \text{if } s \leq n, \\ s-n, & \text{if } s \geq n, \end{cases} \qquad r_2 = \begin{cases} s, & \text{if } s \leq k, \\ k, & \text{if } s > k, \end{cases}$$

and $B = \sum_{i=0}^{k-1} |\beta_j|$. Finally,

$$|e_{n+k}| \le \Gamma\left(HA + \theta_n + hL|\beta_k||e_{n+k}| + hLB \sum_{s=0}^{n+k-1} |e_s|\right)$$

or

$$(1 - hL|\beta_k|\Gamma)|e_{n+k}| \le \Gamma\left(HA + \theta_n + hLB \sum_{s=0}^{n+k-1} |e_s|\right).$$

Supposing now that $h < \dfrac{1}{L|\beta_k|\Gamma}$ and letting $\Gamma^* = \dfrac{1}{1 - hL|\beta_k|\Gamma}$, we get

$$|e_{n+k}| \le \Gamma^* hLB \sum_{s=0}^{n+k-1} |e_s| + \Gamma^*(AH + \theta_n).$$

Corollary 1.6.3 is now applicable and we obtain

$$\begin{aligned}
|e_{n+k}| &\le \Gamma^* AH\, e^{\Gamma^* LB(n+k)h} + \sum_{s=0}^{n+k-1} \Delta\theta_s \exp\left(\sum_{s+1}^{n+k-1} \Gamma^* hLB\right) \\
&= \Gamma^* AH\, e^{\Gamma^* LB(n+k)h} + \sum_{s=0}^{n+k-1} |\tau_s|\, e^{\Gamma^* hLB(n+k-1-s)} \\
&\le \Gamma^* AH\, e^{\Gamma^* LB(n+k)h} + \max_{0\le s\le n+k-1} |\tau_s| \frac{e^{\Gamma^* hLB(n+k)} - 1}{e^{\Gamma^* hLB} - 1}.
\end{aligned}$$

Considering that $(n + k)h \le T$, we get

$$|e_{n+k}| \le \Gamma^* AH\, e^{T\Gamma^* LB} + \max |\tau_s| \frac{e^{\Gamma^* LBT} - 1}{e^{\Gamma^* LB} - 1}.$$

The first term in the right-hand side vanishes in the limit as $n \to \infty$ because we have supposed $\lim_{h\to 0} H = 0$ and $h \le T/n$. The second term becomes $O(h^p)$ because $\tau_s = O(h^{p+1})$ and the denominator is $O(h)$. In view of the hypothesis, we have convergence and the theorem is proved. ■

The previous result holds in the limit when $h \to 0$. It is not very useful in practice because one must use finite integration steps. Then, it is more useful to use the test function (2.7.12). In this case, the equation (2.7.7) becomes

$$\sum_{i=0}^{k} (\alpha_i - h\lambda\beta_i) y_{n+i} = 0. \tag{2.7.14}$$

Setting

$$h\lambda = q \quad \text{and} \quad \pi(z, q) = \rho(z) - q\sigma(z), \tag{2.7.15}$$

equation (2.7.14) becomes

$$\pi(E, q)y_n = 0 \tag{2.7.16}$$

with Re $q \le 0$. The polynomial (2.7.15) is called Dahlquist polynomial.

Definition 2.7.3. The method (ρ, σ) is said to be absolutely stable if the solution $y_n = 0$ of (2.7.16) is stable.

As a simple consequence of Theorem 2.6.2, we have the following result.

Theorem 2.7.4. *The method (ρ, σ) is absolutely stable if $\pi(z, q)$ is a Von Neumann polynomial.*

Definition 2.7.4. The set of values $q \in \mathbb{C}$ for which the method (ρ, σ) is absolutely stable is called the region of absolute stability.

Definition 2.7.5. If the region of absolute stability of the method (ρ, σ) contains the negative complex half-plane, the method is said to be A-stable.

Example. The method defined by

$$y_{n+1} = y_n + h[(1-\theta)f_{n+1} + \theta f_n] \tag{2.7.17}$$

with $0 \le \theta \le 1$ is called θ-method. One has

$$\rho(z) = z - 1, \quad \text{and} \quad \sigma(z) = (1-\theta)z + \theta.$$

It is easily verified that (2.7.9) and (2.7.10) hold. Moreover,

$$\pi(z, q) = (1 - q + q\theta)z - 1 - q\theta$$

whose unique root is $z = \dfrac{1 + q\theta}{1 - q + q\theta}$.

The locus of points satisfying $|z| = 1$ is the circumference with center $\left(\dfrac{1}{1-2\theta}, 0\right)$ and radius $\dfrac{1}{(1-2\theta)^2}$. It is seen that for $0 \le \theta < \frac{1}{2}$, the region of absolute stability is external to the above circle, while for $\frac{1}{2} < \theta \leqslant 1$, such a region is internal. For $\theta = \frac{1}{2}$, the circumference degenerates into the imaginary axis and the region of absolute stability becomes the negative half-plane. As a consequence, we have the following result.

Theorem 2.7.5. *The θ-method is A-stable for $0 \leqslant \theta \leqslant \frac{1}{2}$.*

In the cases $\theta = 0$, $\theta = 1/2$, $\theta = 1$, the θ-method is called respectively implicit Euler method, trapezoidal method and explicit Euler method.

2.8. Boundary Value Problems for Second Order Equations

Here we shall discuss in some detail the most common cases arising in the discretization of the linear second order differential equations. The general problem will be discussed in Chapter 3.

Consider the equation

$$y_{n+1} - 2zy_n + y_{n-1} = 0 \tag{2.8.1}$$

and the three different boundary conditions

$$y_0 = y_N = 0, \tag{2.8.2}$$

$$y_0 = 0, \quad y_{N-1} - zy_N = 0, \tag{2.8.3}$$

$$y_0 = y_N, \quad y_1 = y_{N+1}. \tag{2.8.4}$$

In order to consider these problems, it is very useful to use the general solution of (2.8.1) in the form presented in Example 4 of Section 2.3. First, let us consider the conditions (2.8.2) that give

$$c_1 = 0, \qquad c_2 U_{N-1}(z) = 0, \tag{2.8.5}$$

and

$$y_n = c_2 U_{n-1}(z). \tag{2.8.6}$$

Condition (2.8.5) is satisfied if z is a root of $U_{N-1}(z)$, that is, $z_n = \cos \frac{k\pi}{N}$, $k = 1, 2, \ldots, N-1$, corresponding to which one has $N-1$ solutions $y_n^{(k)} = c_2 U_{n-1}(z_k) = \tilde{c}_2 \sin \frac{kn\pi}{N}$. The parameter $\tilde{c}_2$ is usually chosen by imposing a normalizing condition.

Let us next consider the mixed conditions (2.8.3), which give rise to $c_1 = 0$, $c_2(U_{N-2}(z) - zU_{N-1}(z)) = 0$ and using property (6) of Appendix C, the last equation becomes $T_N(z) = 0$, which is satisfied if $z_k = \cos \frac{2k+1}{N}\frac{\pi}{2}$, $k = 0, 1, \ldots, N-1$, corresponding to which one has solutions

$$y_n^{(k)} = c_2 U_{n-1}(z_k) = c_2^{(k)} \sin \frac{n(2k+1)\pi}{2N}.$$

Finally, the periodic conditions (2.8.4) yield $c_1(1 - T_N(z)) - c_2 U_{N-1}(z) = 0$ and $c_1(T_{N+1}(z) - z) + c_2(-1 + U_N(z)) = 0$.

This system is a homogeneous one and it has nontrivial solutions corresponding to the roots of its determinants which, after some manipulation, using the properties of T_n and U_n given in Appendix C, is equal to $2(1 - T_N(z))$. This expression is zero for $z_k = \cos\frac{2k\pi}{N}$, $k = 0, 1, \ldots, N-1$. One verifies then that the corresponding solutions are

$$y_n^{(k)} = c_1^{(k)} \cos\frac{2k\pi n}{N}, \qquad \text{for } k = 0, \frac{N}{2},$$

$$y_n^{(k)} = c_1^{(k)} \cos\frac{2k\pi n}{N} + c_2^{(k)} \sin\frac{2k\pi n}{N}; \qquad \text{for } k = 1, 2, \ldots, N,\ k \neq \frac{N}{2}.$$

In applications, equation (2.8.1) is usually derived from the discretization of $\frac{d^2y}{dx^2} + \lambda y = 0$ and then the parameter z has the form $z = 1 - \frac{1}{2}\lambda(\Delta x)^2$. The results show that the solution exists only for the values of λ given by $\lambda_k = \frac{2}{(\Delta x)^2}(1 - z_k)$, which are called eigenvalues and the corresponding solutions are called eigenfunctions.

2.9. Problems

2.1. Show that $C_n \det k(n)$ with suitably chosen C_n is a constant of the motion, that is, a relation among the solutions that remains constant for all n.

2.2. Solve the equation $y_{n+2} - 2xy_{n+1} + y_n = 0$, $y_0 = 1$, $y_1 = x$, where x is a real parameter and show that $y_n(x)$ is a polynomial of degree n in x having the coefficient of x^n equal to 2^{n-1}.

2.3. For $|x| \leq 1$, by setting $x = \cos t$, in the previous problem show that

(a) the solution is $y_n = \cos nt$,
(b) the roots of the polynomials $y_n(x)$ are simple and lie in the interval $[-1, 1]$.

The polynomial $y_n(x)$ (usually denoted by $T_n(x)$) are called Chebyshev polynomials and they are of primary importance in Numerical Analysis, see Appendix C.

2.4. Show that the solutions (2.3.8) are linearly independent.

2.5. Suppose we deposit a sum s_0 in a bank that will pay at the end of each year an interest proportional to the initial sum. Determine the amount at the end of rth year.

2.6. In problem (2.5), if the interest paid is proportional to the sum of the deposit, find the amount at the end of rth year.

2.7. Suppose we have to pay an initial debt A with a constant rate R, paying an interest rate i. Determine the rate R if the debt has to be paid in 20 years.

2.8. Equation (2.3.17) deserves a special mention both for its historical importance and for its applications in many different fields.

(a) Take $y_1 = 1$ and $y_2 = 1$. Determine c_1 and c_2 and verify that the sequence obtained is the Fibonacci sequence $1, 1, 2, 3, 5, 8, 13, \ldots$

(b) Show that $\lim_{n\to\infty} \frac{y_{n+1}}{y_n} = \frac{1+\sqrt{5}}{2}$

(This number was called golden section by ancient Greek mathematicians and has been important in the arts through the centuries.)

(c) Show that $\sum_{i=1}^{n} y_i = y_{n+2} - 1$.

2.9. Find the solution of

(a) $y_n(1 + ay_{n-1}) = 1$,

and

(b) $y_n(b + y_{n-1}) = 1$.

$\left(\text{Hint: Set } y_n = \frac{z_n}{z_{n+1}}\right)$.

2.10. The solutions of the previous two problems can be found by using the continuous fractions in the following way (consider, for example, the second case):

$$y_n = \frac{1}{b + y_{n-i}} = \frac{1}{b} + \frac{1}{b + y_{n-2}} = \cfrac{1}{b + \cfrac{1}{b + \cfrac{1}{b + \ldots}}}.$$

When will the fraction converge?

2.11. The best known methods for finding the roots of a single nonlinear equation $f(x) = 0$ are the Newton method and the secant method, defined by

$$x_{n+1} = x_n - \frac{f(x_n)}{f'(x_n)}, \quad \text{and} \quad x_{n+1} = \frac{f(x_n)x_{n-1} - f(x_{n-1})x_n}{f(x_n) - f(x_{n-1})}.$$

If α is the root, denoting the error $|x_n - \alpha|$ with ε_n, show that, for sufficiently smooth f, $\varepsilon_{n+1} = k_n \varepsilon_n^2$ and $\varepsilon_{n+1} = k'_n \varepsilon_n \varepsilon_{n-1}$, respectively k_n and k'_n uniformly bounded. The order of convergence of an iterative method is defined by $p = \lim_{n\to\infty} \frac{\log \varepsilon_{n+1}}{\log \varepsilon_n}$. Show that in the first case $p = 2$, while in the second case $p = \frac{1+\sqrt{5}}{2}$.

2.12. Solve the difference equation $z_{n+1} + z_n = -(n+1)$.

2.13. (*Bernoulli Method*). The solution of the linear difference equation has been obtained considering the polynomial $p(z)$. It also happens that for finding the roots of a polynomial it is useful to consider a linear difference equation. In fact, supposing that the roots $z_1, z_2, \ldots, z_k$ are all simple and that $|z_1| > |z_2| > \ldots$ from (2.3.7) we have

$$y_n = \sum_{i=1}^{k} c_i z_i^n = c_1 z_1^n \left(1 + \sum_{i=2}^{k} \frac{c_i}{c_1}\left(\frac{z_i}{z_1}\right)^n\right)$$

from which it follows

$$\lim_{n\to\infty} \frac{y_{n+1}}{y_n} = z_1.$$

Solving then the difference equation recursively, the ratio of two successive values of the solution will give an approximation of the first root.

(a) How can we approximate the root of minimum modulus?
(b) What happens if $z_1 \simeq z_2$?
(c) How do we choose the initial conditions in order to avoid the effect of multiple zeros of the characteristic polynomial?

2.14 Suppose one has to perform the sum $S_n = \sum_{i=1}^{n} a_i$ by using the following algorithm: $S_0 = 0$, $S_{i+1} = S_i + a_{i+1}$. If one performs the sum not using the real numbers, but approximation of them (floating point numbers), that is, instead of the number $a = m \cdot 10^q$, where $0.1 \le \bar{m} < 1$, one uses the number $\bar{a} = \bar{m} \cdot 10^q$, where $0.1 \le \bar{m} < 1$ but $\bar{m}$ has only t digits, this implies that $|a - \bar{a}| < |m - \bar{m}| \cdot 10^q \le 10^{q-t}$. Study the behavior of the errors, considering that $\bar{s}_i = (\bar{s}_{i-1} + \bar{a}_i)(1 + 10^{-t})$.

2.15. If

$$A_n = \begin{pmatrix} -2 & 1 & 0 & \cdots & & 0 \\ 1 & -2 & 1 & \ddots & & \vdots \\ 0 & \ddots & \ddots & \ddots & & 0 \\ \vdots & & \ddots & \ddots & \ddots & 1 \\ 0 & \cdots & & 0 & 1 & -2 \end{pmatrix}_{n \times n},$$

Show that $D_n = \det A_n$ satisfies the equation $D_{n+2} + 2D_{n+1} + D_n = 0$ and that $D_n = (-1)^n(n+1)$.

2.16. If

$$A_n = \begin{pmatrix} 2a & b & 0 & \cdots & & 0 \\ c & \ddots & \ddots & \ddots & & \vdots \\ 0 & \ddots & \ddots & \ddots & & 0 \\ \vdots & & \ddots & \ddots & \ddots & b \\ 0 & \cdots & & 0 & c & 2a \end{pmatrix}_{n \times n},$$

and $D_n(\lambda) = \det(A_n - \lambda I)$, find the eigenvalues of A_n in the case $a^2 \leq bc$.

2.17. If $\{y_n\}$ has generating function $f(z)$, for $|z| < \boldsymbol{R}$, show that $y_n/(n+1)$ has generating function $1/z \int_0^z f(t)\,dt$ in the same region. (Hint: integrate term by term.)

2.18. Prove (2.7.9). (Hint: expand in Taylor series starting from $y(t_n)$ and equate to zero the coefficients of h^0 and h.)

2.19. Show that the solution of $y_{n+2} - \frac{3}{2} y_{n+1} + \frac{1}{2} y_n = \frac{1}{n+1}$, $y_0 = 0$, $y_1 = 0$, is unbounded.

2.20. Prove that if the method (ρ, σ) is convergent (for all f satisfying the hypothesis stated in the text), then it is 0-stable. (Hint: take $f = 0$.)

2.21. With the hypothesis of Problem 2.20, prove that the method is consistent.

2.22. Suppose that the relation (2.7.8) is given by $\rho(E)y_n - h\sigma(E)f_n = \varepsilon_n$, where ε_n is a small bounded quantity but not infinitesimal with respect to h. (This happens in practice when we solve the difference equation on the computer.) Using a similar procedure used in the text prove that $|e_{n+k}| \leq E_1 + E_2$, where E_1 and E_2 have respectively a zero and a pole for $h = 0$. Deduce for this that it is not convenient in practice to use h arbitrarily small.

2.23. The linear multistep method $z_{n+1} - 2z_{n+1} + z_n = h[f_{n+1} - f_n]$ is consistent. It is not 0-stable. Show, however, that for the some set of initial conditions the solutions of $\rho(E)z_n = 0$ remain bounded. (Hint: try with constant initial conditions.)

2.24. Find the region of absolute stability for the following methods:

(a) $y_{n+2} - y_n = 2hf_{n+1}$, (midpoint),

(b) $y_{n+2} - y_n = \frac{h}{3}(f_{n+2} + 4f_{n+1} + f_n)$, (Simpson rule).

2.25. Find the solution of the boundary value problem $y_{n+2} - 2zy_{n+1} + y_n = 0$, $y_1 - zy_0 = 0$; $y_{N-1} - zy_N = 0$.

2.10. Notes

The material of Sections 2.1 to 2.4 is classical and can be found in many classical books. Theorem 2.1.8 is essentially a compact form of a result in Clenshaw [25] and Luke [99], see also Section 5.4. The content of Section 2.5 is also classical but the notation is adapted from Henrici's book [13]. Section 2.6 consists of a collection of results that are scattered in many publications, essentially dealing with Numerical Analysis. The material of Section 2.7 is based on Dahlquist's papers starting from the fundamental one [33] (see also references given in Chapter 6). The books of Lambert [92] and Gear [54] give more detailed arguments on the subject. Theorem 2.7.3 can be found in Henrici [75]. Section 2.8 deals with material that can be found either in some books on numerical analysis (see for example [150]) or in some books on difference equations, see Fort [49]. The classical reference is the book of Atkinson [9] where both the continuous and discrete cases are treated as well as many applications.

CHAPTER 3 Linear Systems of Difference Equations

3.0. Introduction

In this chapter, we shall treat systems of linear difference equations. Some results discussed in Chapter 2 are presented here in an elegant form in terms of matrix theory. After investigating the basic theory, method of variation of constants and higher order systems in Sections 3.1 to 3.3, we shall consider the case of periodic solutions in Section 3.4. Boundary value problems are dealt with in Section 3.5, where the classical theory of Poincaré is also included. The elements of matrix theory that are necessary for this chapter may be found in Appendix A. Some useful problems are given in Section 3.6.

3.1. Basic Theory

Let $A(n)$ be an $s \times s$ matrix whose elements $a_{ij}(n)$ are real or complex functions defined on $\boldsymbol{N}^+_{n_0}$ and $y_n \in \boldsymbol{R}^s$ (or $\boldsymbol{C}^s$) with components that are functions defined on the same set $\boldsymbol{N}^+_{n_0}$.

A linear equation

$$y_{n+1} = A(n)y_n + b_n, \tag{3.1.1}$$

where $b_n \in \boldsymbol{R}^s$ and y_{n_0} is a given vector is said to be a nonhomogeneous linear difference equation. The corresponding homogeneous linear difference equation is

$$y_{n+1} = A(n)y_n. \tag{3.1.2}$$

When an initial vector y_{n_0} is assigned, both (3.1.1) and (3.1.2) determine the solution uniquely on the set $N_{n_0}^+$ as can be seen easily as induction. For example, it follows from (3.1.2) that the solution takes the form

$$y_n = \left[\prod_{i=n_0}^{n-1} A(i)\right] y_{n_0}, \tag{3.1.3}$$

from which follows the uniqueness of solution passing through y_{n_0}, because $\prod_{i=n_0}^{n-1} A(i)$ is uniquely defined for all n. Sometimes, in order to avoid confusion, we shall denote by $y(n, n_0, y_{n_0})$ the solution of (3.1.1) or (3.1.2) having y_{n_0} as initial vector.

Let us now consider the space S of solutions of (3.1.2). It is a linear space since by taking any two solutions of (3.1.2), we can show that any linear combination of them is a solution of the same equation.

Let $E_1, E_2, \ldots, E_s$ be the unit vectors of $\boldsymbol{R}^s$ and $y(n, n_0, E_i)$, $i = 1, 2, \ldots, s$, the s solutions having E_i as initial vectors.

Lemma 3.1.1. *Any element of $\boldsymbol{S}$ can be expressed as a linear combination of $y(n, n_0, E_i)$, $i = 1, 2, \ldots, s$.*

Proof. Let $y(n, n_0, c)$ be a solution of (3.1.2) with $y_{n_0} = c \in \boldsymbol{R}^s$. From the linearity of $\boldsymbol{S}$ and from $c = \sum_{i=1}^{s} c_i E_i$, it follows that the vector

$$z_n = \sum_{i=1}^{s} c_i y(n, n_0, E_i)$$

satisfies (3.1.2) and has c as initial vector. Then, by uniqueness, z_n must coincide with $y(n, n_0, c)$. ∎

Definition 3.1.1. Let $f_i(n)$, $i = 1, 2, \ldots, s$, be vector valued functions defined on $N_{n_0}^+$. They are linearly dependent if there exists constants a_i, $i = 1, 2, \ldots, s$ not all zero such that $\sum_{i=1}^{s} a_i f_i(n) = 0$, for all $n \geq n_0$.

Definition 3.1.2. The vectors $f_i(n)$, $i = 1, 2, \ldots, s$, are linearly independent if they are not linearly dependent.

Let us define the matrix $K(n) = (f_1(n), f_2(n), \ldots, f_s(n))$ whose columns are the vectors $f_i(n)$. Also let a be the vector $(a_1, a_2, \ldots, a_s)^T$.

Theorem 3.1.1. *If there exists an $\bar{n} \in N_{n_0}^+$ such that* $\det K(\bar{n}) \neq 0$ *then the vectors $f_i(n)$, $i = 1, 2, \ldots, s$ are linearly independent.*

Proof. Suppose that for $n \geq n_0$

$$K(n)a = \sum_{i=1}^{s} a_i f_i(n) = 0.$$

Since $\det K(\bar{n}) \neq 0$, it follows that $a = 0$ and the functions $f_i(n)$ are not linearly dependent. ■

Theorem 3.1.2. *If $f_i(n)$, $i = 1, 2, \ldots, s$, are solutions of* (3.1.2) *with* $\det A(n) \neq 0$ *for* $n \in \mathbf{N}_{n_0}^+$, *and if* $\det K(n_0) \neq 0$, *then* $\det K(n) \neq 0$ for all $n \in \mathbf{N}_{n_0}^+$.

Proof. For $n \geq n_0$,

$$\begin{aligned} \det K(n+1) &= \det(f_1(n+1), f_2(n+1), \ldots, f_s(n+1)) \\ &= \det A(n) \det K(n), \end{aligned} \tag{3.1.4}$$

from which it follows that

$$\det K(n) = \left(\sum_{i=n_0}^{n-1} \det A(i) \right) \det K(n_0). \quad ■ \tag{3.1.5}$$

Corollary 3.1.1. *The solutions $y(n, n_0, E_i)$, $i = 1, 2, \ldots, s$, of* (3.1.2) *with* $\det A(n) \neq 0$ *for* $n \geq n_0$, *are linearly independent.*

Proof. In this case $\det K(n_0) = I$, the identity matrix and by Theorem 3.1.1, the result follows. ■

Corollary 3.1.2. *If the columns of $K(n)$ are linearly independent solutions of* (3.1.2) *with* $\det A(n) \neq 0$, *then* $\det K(n) \neq 0$ *for all* $n \geq n_0$.

Proof. The proof follows from the fact that there exists an $\bar{n}$ at which $\det K(\bar{n}) \neq 0$ and the relation (3.1.4). ■

The matrix $K(n)$, when its columns are solution of (3.1.2) is called Casorati matrix or fundamental matrix. We shall reserve the name of fundamental matrix for a slightly different matrix, and call $K(n)$ the Casorati matrix. Its determinant is called Casoratean and plays the same role as the Wronskian in the continuous case.

The Casorati matrix satisfies the equation

$$K(n+1) = A(n)K(n). \tag{3.1.6}$$

Theorem 3.1.3. *The space S of all solutions of* (3.1.2) *is a linear space of dimension s.*

The proof is an easy consequence of Lemma 3.1.1 and Corollary 3.1.1.

Definition 3.1.3. Given s linearly independent solutions of (3.1.2), and a vector $c \in \boldsymbol{R}^s$ of arbitrary components, the vector valued function $y_n = K(n)c$ is said to be the general solution of (3.1.2).

Fixing the initial condition y_{n_0}, it follows from definition 3.1.3 that $c = K^{-1}(n_0)y_{n_0}$ and

$$y(n, n_0, y_{n_0}) = K(n)K^{-1}(n_0)y_{n_0} \tag{3.1.7}$$

and in general, for $s \in \boldsymbol{N}^+_{n_0}$, $y_s = c$,

$$y(n, s, c) = K(n)K^{-1}(s)c. \tag{3.1.8}$$

The matrix

$$\Phi(n, s) = K(n)K^{-1}(s) \tag{3.1.9}$$

satisfies the same equation as $K(n)$, i.e., $\Phi(n+1, s) = A(n)\Phi(n, s)$. Moreover, $\Phi(n, n) = I$ for all $n \geq n_0$. We shall call Φ the fundamental matrix. In terms of the fundamental matrix, (3.1.7) can be written as $y(n, n_0, y_{n_0}) = \Phi(n, n_0)y_{n_0}$. Other properties of the matrix Φ are (i) $\Phi(n, s)\Phi(s, t) = \Phi(n, t)$ and (ii) if $\Phi^{-1}(n, s)$ exists then

$$\Phi^{-1}(n, s) = \Phi(s, n). \tag{3.1.10}$$

The relation (3.1.10) allows us to define $\Phi(s, n)$, for $s < n$.

Let us now consider the nonhomogeneous equation (3.1.1).

Lemma 3.1.2. *The difference between any two solutions y_n and $\bar{y}_n$ of* (3.1.1) *is a solution of* (3.1.2).

Proof. From the fact that

$$y_{n+i} = A(n)y_n + b_n, \qquad \bar{y}_{n+1} = A(n)\bar{y}_{n+1} + b_n,$$

one obtains

$$y_{n+1} - \bar{y}_{n+1} = A(n)(y_n - \bar{y}_{n+1}),$$

which proves the lemma. ∎

Theorem 3.1.4. *Every solution of* (3.1.1) *can be written in the form $y_n = \bar{y}_n + \Phi(n, n_0)c$, where $\bar{y}_n$ is a particular solution of* (3.1.1) *and $\Phi(n, n_0)$ is the fundamental matrix of the homogeneous equation* (3.1.2).

Proof. From Lemma 3.1.2, $y_n - \bar{y}_n \in \boldsymbol{S}$ and an element in this space can be written in the form $\Phi(n, n_0)c$.

If the matrix A is independent of n, the fundamental matrix simplifies because $\Phi(n, n_0) = \Phi(n - n_0, 0)$. ∎

3.2. Method of Variation of Constants

From the general solution of (3.1.2), it is possible to obtain the general solution of (3.1.1). The general solution of (3.1.2) is given by $y(n, n_0, c) = \Phi(n, n_0)c$. Let c be a function defined on $N^+_{n_0}$ and let us impose the condition that $y(n, n_0, c_n)$ satisfy (3.1.1). We then have

$$y(n+1, n_0, c_{n+1}) = \Phi(n+1, n_0)c_{n+1} = A(n)\Phi(n, n_0)c_{n+1}$$
$$= A(n)\Phi(n, n_0)c_n + b_n,$$

from which, supposing that $\det A(n) \neq 0$ for all $n \geq n_0$, we get

$$c_{n+1} = c_n + \Phi(n_0, n+1)b_n.$$

The solution of the above equation is

$$c_n = c_{n_0} + \sum_{j=n_0}^{n-1} \Phi(n_0, j+1)b_j.$$

The solution of (3.1.1) can now be written as

$$y(n, n_0, c_{n_0}) = \Phi(n, n_0)c_{n_0} + \sum_{j=n_0}^{n-1} \Phi(n, n_0)\Phi(n_0, j+1)b_j$$
$$= \Phi(n, n_0)c_{n_0} + \sum_{j=n_0}^{n-1} \Phi(n, j+1)b_j,$$

from which, setting $c_n = y_{n_0}$, we have

$$y(n, n_0, y_{n_0}) = \Phi(n, n_0)y_{n_0} + \sum_{j=n_0}^{n-1} \Phi(n, j+1)b_j. \tag{3.2.1}$$

By comparing (3.1.5) and (3.1.9), it follows that $\Phi(n, n_0) = \prod_{i=n_0}^{n-1} A(i)$, where $\prod_{i=n_0}^{n_0-1} A(i) = I$. We can rewrite (3.2.1) in the form

$$y(n, n_0, y_{n_0}) = \left(\prod_{i=n_0}^{n-1} A(i)\right)y_{n_0} + \sum_{j=n_0}^{n-1}\left(\prod_{s=j+1}^{n-1} A(s)\right)b_j. \tag{3.2.2}$$

In the case where A is a constant matrix $\Phi(n, n_0) = A^{n-n_0}$ and, of course, $\Phi(n, n_0) = \Phi(n - n_0, 0)$. The equation (3.2.2) reduces to

$$y(n, n_0, y_{n_0}) = A^{n-n_0}y_{n_0} + \sum_{j=n_0}^{n-1} A^{n-j-1}b_j. \tag{3.2.3}$$

Let us now consider the case where $A(n)$ as well as b_n are defined on $N^{\pm}$.

Theorem 3.2.1. *Suppose that* $\sum_{j=-\infty}^{n} \|K^{-1}(j+1)\| < +\infty$ *and* $\|b_j\| < b$, $b \in R^+$, $j \in N^{\pm}$. *Then,*

$$y_n = \sum_{s=0}^{\infty} K(n)K^{-1}(n-s)b_{n-s-1} \tag{3.2.4}$$

is a solution of (3.1.1).

Proof. For $m \in \mathbf{N}^{\pm}$ consider the solution, corresponding to $y_m = 0$,

$$y(n, m, 0) = \sum_{j=m}^{n-1} K(n)K^{-1}(j+1)b_j$$

and the sequence $y(n, m-1, 0)$, $y(n, m-2, 0), \ldots$.

This sequence is a Cauchy sequence since, for $r > 0$, $\varepsilon > 0$ and m_1 chosen such that $\sum_{j=m_1-r}^{m_1} \|K^{-1}(j+1)\| < \varepsilon$ and

$$\begin{aligned}\|y(n, m_1 - r, 0) - y(n, m_1, 0)\| &= \left\| \sum_{j=m_1-r}^{m_1} K(n)K^{-1}(j+1)b_j \right\| \\ &\leq \|K(n)\| b\varepsilon.\end{aligned}$$

It follows that the sequence will converge as $m \to -\infty$. Let y_n be the limit given by

$$y_n = \sum_{j=-\infty}^{n-1} K(n)K^{-1}(j+1)b_j,$$

which is again a solution of (3.1.1). By setting $s = n - j - 1$, we obtain

$$y_n = \sum_{s=0}^{\infty} K(n)K^{-1}(n-s)b_{n-s-1}.$$

In the case of constant coefficients this solution takes the form

$$y_n = \sum_{s=0}^{\infty} A^s b_{n-s-1}, \tag{3.2.5}$$

which exists if the eigenvalues of A are inside the unit circle. ■

Let us close this section by giving the solution in a form that corresponds to the one given using the formal series in the scalar case.

Let

$$y_{n+1} = Ay_n + b_n. \tag{3.2.6}$$

By multiplying with z^n, with $z \in \mathbf{C}$, and summing formally from zero to infinity, one has

$$\frac{1}{z} \sum_{n=0}^{\infty} y_{n+1} z^{n+1} = A \sum_{n=0}^{\infty} y_n z^n + \sum_{n=0}^{\infty} b_n z^n.$$

Letting

$$Y(z) = \sum_{n=0}^{\infty} y_n z^n, \qquad B(z) = \sum_{n=0}^{\infty} b_n z^n$$

and substituting, one obtains

$$z^{-1}(Y(z)-y_0)=AY(z)+B(z)$$

from which

$$(I-zA)Y(z)=y_0+zB(z)$$

and

$$Y(z)=z^{-1}(z^{-1}I-A)^{-1}(y_0+zB(z)). \tag{3.2.7}$$

When the formal series is convergent, the previous formula furnishes the solutions as the coefficient vectors of $Y(z)$. The matrix $\boldsymbol{R}(z^{-1},A)=(z^{-1}I-A)^{-1}$ is called resolvent of A (see A.3). Its properties reflect the properties of the solution y_n.

3.3. Systems Representing High Order Equations

Any k^{th}-order scalar linear difference equation

$$y_{n+k}+p_1(n)y_{n+k-1}+\ldots+p_k(n)y_n=g_n \tag{3.3.1}$$

can be written as a first order system in $\boldsymbol{R}^k$, by introducing the vectors

$$Y_n=\begin{pmatrix} y_n \\ y_{n+1} \\ \vdots \\ y_{n+k-1}\end{pmatrix},\qquad Y_0=\begin{pmatrix} y_0 \\ y_1 \\ \vdots \\ y_{k-1}\end{pmatrix},\qquad G_n=\begin{pmatrix} 0 \\ 0 \\ \vdots \\ g_n\end{pmatrix} \tag{3.3.2}$$

and the matrix

$$A(n)=\begin{pmatrix} 0 & 1 & 0 & & \ldots & 0 \\ 0 & 0 & 1 & 0 & \ldots & 0 \\ \vdots & & & & & \vdots \\ & & & & & 0 \\ 0 & & \ldots 0 & & & 1 \\ -p_k(n) & & -p_{k-1}(n) & & \ldots & -p_1(n)\end{pmatrix}. \tag{3.3.3}$$

Using this notation equation, (3.3.1) becomes

$$Y_{n+1}=A(n)Y_n+G_n \tag{3.3.4}$$

where Y_0 is the initial condition.

The matrix $A(n)$ is called the companion (or Frobenius) matrix and some of its interesting properties that characterize the solution of (3.3.4) are given:

(i) The determinant of $A(n)-\lambda I$ is the polynomial $(-1)^k\ (\lambda^k+p_1(n)\lambda^{k-1}+\ldots+p_k(n))$. When A is independent of n this polynomial coincides with the characteristic polynomial;

(ii) $\det A(n) = (-1)^k p_k(n)$ and is nonsingular if (3.3.1) is really a kth order equation;

(iii) There are no semisimple eigenvalues of $A(n)$ (see Appendix A). This implies that both the algebraic and geometric multiplicity of the eigenvalues of A coincide. This property is important in determining the qualitative behavior of the solutions;

(iv) When A is independent of n and has simple eigenvalues $z_1, z_2, \ldots, z_s$, it can be diagonalized by the similarity transformation $A = VDV^{-1}$, where V is the Vandermonde matrix $V(z_1, z_2, \ldots, z_s)$ and $D = \text{diag}(z_1, z_2, \ldots, z_s)$.

The solution of (3.3.4) is deduced by (3.2.1), which in the present notation becomes

$$Y_n = \Phi(n, n_0)Y_0 + \sum_{j=n_0}^{n-1} \Phi(n, j+1)G_j.$$

The fundamental matrix $\Phi(n, n_0)$ is given by

$$\Phi(n, n_0) = K(n)K^{-1}(n_0),$$

where the Casorati matrix $K(n)$ is given in terms of k independent solutions $f_1(n), f_2(n), \ldots, f_k(n)$ of the homogeneous equation corresponding to (3.3.4), i.e.,

$$K(n) = \begin{pmatrix} f_1(n) & f_2(n) & \cdots & f_k(n) \\ f_1(n+1) & f_2(n+1) & & \vdots \\ \vdots & \vdots & & \vdots \\ f_1(n+k-1) & f_2(n+k-1) & \cdots & f_k(n+k-1) \end{pmatrix}.$$

The solution Y_n of (3.3.4) has redundant information concerning the solution of (3.3.1). It is enough to consider any component of Y_n for $n \geq n_0 + k$ to get the solution of (3.3.1). For example, if we take the case $Y_0 = 0$, from (3.2.1) we have

$$Y_n = \sum_{j=n_0}^{n-1} \Phi(n, j+1)G_j, \tag{3.3.5}$$

where, by (3.1.9), $\Phi(n, j+1) = K(n)K^{-1}(j+1)$. To obtain the solution $y(n+k, n_0, 0)$ of (3.3.1), it is sufficient to consider the last component of the vector Y_{n+1} and we get

$$\begin{aligned} y_{n+k} &= \sum_{j=n_0}^{n} E_k^T \Phi(n+1, j+1)G_j \\ &\equiv \sum_{j=n_0}^{n} E_k^T K(n+1)K^{-1}(j+1)E_k g_j, \end{aligned} \tag{3.3.6}$$

where $E_k = (0, 0, \ldots, 0, 1)^T$.

Introducing the functions

$$H(n+k,j) = E_k^T(n+1)K^{-1}(j+1)E_k, \tag{3.3.7}$$

the solution (3.3.6) can be written as

$$y_{n+k} = \sum_{j=n_0}^{n} H(n+k,j)g_j. \tag{3.3.8}$$

The function $H(n+k,j)$, which is called the one-sided Green's function has some interesting properties. For example, it follows easily from (3.3.7) that

$$H(n+k,n) = 1. \tag{3.3.9}$$

In order to obtain additional properties, let us consider the identity

$$I = \sum_{i=1}^{k} E_i E_i^T, \tag{3.3.10}$$

where E_i are the unit vectors in $\boldsymbol{R}^k$ and I is the identity matrix. From (3.3.7), one has

$$H(n+k,j) = \sum_{i=1}^{k} E_k^T K(n+1) E_i E_i^T K^{-1}(j+1)E_k, \tag{3.3.11}$$

which represents the sum of the products of the elements of the last row of $k(n+1)$ and the elements of the last column of $k^{-1}(j+1)$. By observing that the elements in the last column of the matrix $K^{-1}(j+1)$ are the cofactors of the elements of the last row of the matrix $K(j+1)$, it follows that

$$H(n+k,j) = \frac{1}{\det K(j+1)} \times \det \begin{pmatrix} f_1(j+1) & \cdots & \cdots & f_k(j+1) \\ f_1(j+2) & \cdots & \cdots & f_k(j+2) \\ \cdots & \cdots & \cdots & \cdots \\ \cdots & \cdots & \cdots & \cdots \\ f_1(j+k-1) & & & f_k(j+k-1) \\ f_1(n+k) & f_2(n+k) & \cdots & f_k(n+k) \end{pmatrix}. \tag{3.3.12}$$

As a consequence, one finds

$$H(n+k,n) = 1, \tag{3.3.13}$$

$$H(n+k-i,n) = 0, \qquad i = 1, 2, \ldots, k-1, \tag{3.3.14}$$

and

$$H(n+k,n+k) = (-1)^{k-1}\frac{\det K(n+k)}{\det K(n+k+1)} = -\frac{1}{p_k(n+k)}. \tag{3.3.15}$$

Proposition 3.3.1. *The function $H(n,j)$, for fixed j, satisfies the homogeneous equation associated with* (3.3.1), *that is,*

$$\sum_{i=0}^{k} p_i(n)H(n+k-i,j) = 0.$$

Proof. It follows easily from (3.3.12) and the properties of the determinant. ■

The solution (3.3.8) can also be written as

$$y_n = \sum_{j=n_0}^{n-k} H(n,j)g_j, \qquad n \geq n_0, \tag{3.3.16}$$

with the usual convention $\sum_{n_0}^{s<n_0} = 0$.

For the case of arbitrary initial conditions together with equation (3.3.1), one can proceed in similar way. From the solution

$$Y_{n+1} = K(n+1)K^{-1}(n_0)y_0 + \sum_{j=n_0}^{n} K(n+1)K^{-1}(j+1)G_j,$$

by taking the kth component we have

$$y_{n+k} = E_k^T K(n+1)K^{-1}(n_0)y_0 + \sum_{j=n_0}^{n} H(n+k,j)g_j.$$

In the case of constant coefficients, the expression for $H(n,j)$ can be simplified. Suppose that the roots z_i of the characteristic polynomial are all distinct. We then have, from (3.3.11),

$$H(n+k,j) = \frac{\prod_{i=1}^{k} z_i^{j+1}}{\prod_{i=1}^{k} z_i^{j+1} \det K(0)} \det \begin{pmatrix} 1 & \cdots & 1 \\ z_1 & \cdots & z_k \\ & \cdots & \\ z_1^{k-2} & \cdots & z_k^{k-2} \\ z_1^{(n-j)+k-1} & \cdots & z_k^{(n-j)+k-1} \end{pmatrix}$$

$$= \frac{\sum_{i=1}^{k} z_i^{n-j+k-1} V_i(z_1,\ldots,z_k)}{\det K(0)} = \sum_{i=1}^{k} \frac{z_i^{n-j+k-1}}{p'(z_i)}, \tag{3.3.17}$$

where $V_i(z_1,\ldots,z_k)$ are the cofactors of the ith elements of the last row and $p'(z_i)$ is the derivative of the characteristic polynomial evaluated at z_i.

In this case, as can be expected, one has $H(n+k,j) = H(n+k-j,0)$. By denoting with $H(n+k-j)$ the function $H(n+k-j,0)$, the solution of the equation

$$\sum_{i=1}^{k} p_i y_{n+k-i} = g_n$$

such that $y_i = 0$, $i = 0, 1, \ldots, k-1$, is given by

$$y_n = \sum_{j=0}^{n-k} H(n-j) g_j. \tag{3.3.18}$$

The properties (3.3.12), (3.3.13), (3.3.14) reduce to $H(k) = 1$, $H(k-s) = 0$, $s = 1, \ldots, k-1$, and $H(0) = -\frac{1}{p_k}$, respectively.

We shall now state two classical theorems on the growth of solutions of (3.3.1) with $g_n = 0$.

Theorem 3.3.1. (*Poincaré*). *If* $\lim_{n\to\infty} p_i(n) = p_i$, $i = 1, 2, \ldots, k$, *and if the roots of*

$$\sum_{i-0}^{k} p_i \lambda^{k-i} = 0, \qquad p_0 = 1 \tag{3.3.19}$$

have distinct moduli, then for every solution y_n,

$$\lim_{n\to\infty} \frac{y_{n+1}}{y_n} = \lambda_s,$$

where λ_s *is a solution of* (3.3.19).

Proof. Let $p_i(n) = p_i + \eta_i(n)$ where, by hypothesis, $\eta_i \to 0$ as $n \to \infty$. The matrix $A(n)$ can be split as $A(n) = A + E_k \eta^T(n)$, where

$$A = \begin{pmatrix} 0 & 1 & 0 & \ldots & 0 \\ 0 & 0 & 1 & 0 & \vdots \\ \vdots & & & & \vdots \\ 0 & & \ldots & 0 & 1 \\ -p_k & -p_{k-1} & & \ldots & -p_1 \end{pmatrix}$$

$\eta^T(n) = (-\eta_k(n), \eta_{k-1}(n), \ldots, -\eta_1(n))$ and $E_k^T = (0, 0, \ldots, 1)$. Equation (3.3.4) then becomes $Y_{n+1} = AY_n + E_k \eta^T(n) Y_n$. Now $A = V \Lambda V^{-1}$, where V is the Vandermonde matrix made up of the eigenvalues $\lambda_1, \lambda_2, \ldots, \lambda_k$ of A (which are the roots of (3.3.19)) and $\Lambda = \text{diag}(\lambda_1, \lambda_2, \ldots, \lambda_k)$. We suppose that $|\lambda_1| < |\lambda_2| < \ldots < |\lambda_k|$. Changing variables $u(n) = V^{-1} Y_n$ and letting $\Gamma_n = V^{-1} E_k \eta^T(n) V$, we get $u(n+1) = \Lambda u(n) + \Gamma_n u(n)$. The elements of $\Gamma(n)$, being linear combinations of $\eta_i(n)$, tend to zero as $n \to \infty$. This implies that for any matrix norm, we have $\|\Gamma_n\| \to 0$. Suppose now that

$\max_{1 \le i \le k} |u_i(n)| = |u_s(n)|$. The index s will depend on n. We shall show that an n_0 can be chosen such that for $n \ge n_0$, the function $s(n)$ is not decreasing. In fact, we know that for $i < j$, $\dfrac{|\lambda_i|}{|\lambda_j|} < 1$. Take $\varepsilon > 0$ small enough such that $\dfrac{|\lambda_i| + \varepsilon}{|\lambda_j| - \varepsilon} < 1$ and choose n_0 so that for $n \ge n_0$, $\|\Gamma_n\|_\infty < \varepsilon$. Setting $s(n+1) = j$, it follows that

$$|u_s(n+1)| \ge |\lambda_s||u_s(n)| - \|\Gamma_n\|_\infty |u_s(n)| = (|\lambda_s| - \varepsilon)|u_s(n)|$$

and

$$|u_j(n+1)| \le |\lambda_j||u_j(n)| + \varepsilon|u_s(n)| \le (|\lambda_j| + \varepsilon)|u_s(n)|$$
$$|u_j(n+1)| \ge |\lambda_j||u_j(n)| - \varepsilon|u_s(n)|.$$

Consequently, if $s(n+1) \equiv j$ were less than $s(n)$, the following would be true:

$$\frac{|u_j(n+1)|}{|u_s(n+1)|} \le \frac{|\lambda_j| + \varepsilon}{|\lambda_2| - \varepsilon} < 1,$$

which is in contradiction with the definition of j. For $n > \mathbf{N}$ suitably chosen, the function $s(n)$ will then assume a fixed value less or equal to k. We shall show now that the ratios

$$\frac{|u_j(n)|}{|u_s(n)|}, \qquad j \ne s \tag{3.3.20}$$

tend to zero. In fact we know that for $n > \mathbf{N}$, $\dfrac{|u_j(n)|}{|u_s(n)|} \le \alpha < 1$. This means that α is an upper limit for (3.3.20). We extract for $n \ge \mathbf{N}$ a subsequence $n_1, n_2, \ldots,$ for which (3.3.20) converges to α. Suppose first that $j > s$. Then

$$\frac{|u_j(n_p+1)|}{|u_s(n_p+1)|} \ge \frac{|\lambda_j|\dfrac{|u_j(n_p)|}{|u_s(n_p)|} - \varepsilon}{|\lambda_s| + \varepsilon}.$$

We take the limit of subsequence, obtaining a lower limit

$$\lim_{p\to\infty} \frac{|u_j(n_p+1)|}{|u_s(n_p+1)|} \ge \frac{|\lambda_j|\alpha - \varepsilon}{|\lambda_s| + \varepsilon}.$$

This implies

$$\frac{|\lambda_j|\alpha - \varepsilon}{|\lambda_s| + \varepsilon} \le \lim_{p\to\infty} \frac{|u_j(n_p+1)|}{|u_s(n_p+1)|} \le \overline{\lim_{p\to\infty}} \frac{|u_j(n_p+1)|}{|u_s(n_p+1)|} = \alpha$$

for arbitrary small ε. Since $\dfrac{|\lambda_j|}{|\lambda_s|} > 1$, the previous relation holds only for $\alpha = 0$. In the case $j < s$, similar arguments, starting from

$$\frac{|u_j(n_p+1)|}{|u_s(n_p+1)|} \le \frac{|\lambda_j| \dfrac{|u_j(n_p)|}{|u_s(n_p)|} + \varepsilon}{|\lambda_s| - \varepsilon}$$

lead to the same conclusion that $\alpha = 0$.

Consider now the original solution y_n. We have, considering the first two rows of $Y_n = Vu(n)$:

$$y_n = \sum_{i \ne s} u_i(n) + u_s(n) = u_s(n)\left(1 + \sum_{i \ne s} \frac{u_i(n)}{u_s(n)}\right),$$

and

$$y_{n+1} = \sum_{i \ne s} \lambda_i u_i(n) + \lambda_s u_s(n) = \lambda_s u_s(n)\left(1 + \sum_{i \ne s} \frac{\lambda_i u_i(n)}{\lambda_s u_s(n)}\right).$$

One easily verifies, by using the previous results, that

$$\lim_{n \to \infty} \frac{y_{n+1}}{y_n} = \lambda_s. \qquad \blacksquare$$

We shall now state, without proof, the following theorem due to Perron, which improves the previous one.

Theorem 3.3.2. (*Perron*). *Suppose that the hypotheses of Theorem* 3.3.1 *are verified, and moreover* $p_k(n) \ne 0$ *for* $n \ge n_0$. *Then* k *solutions* $f_1, f_2, \ldots, f_k$ *can be found such that*

$$\lim_{n \to \infty} \frac{f_i(n+1)}{f_i(n)} = \lambda_i, \qquad i = 1, 2, \ldots, k.$$

3.4. *Periodic Solutions*

Let N be a positive integer greater than 1, $A(n)$ real nonsingular $s \times s$ matrices and b_n vectors of $\boldsymbol{R}^s$. Suppose that $A(n)$ and b_n are periodic of period N. That is $A(n+N) = A(n)$ and $b_{n+N} = b_n$. A period solution of period N is a solution for which $y_{n+N} = y_n$.

The trivial example of periodic solution of the homogeneous equation (3.1.2) is $y_n = 0$. For the nonhomogeneous equation (3.1.1), if there exists a constant vector $\bar{y}$ such that $(A(n) - I)\bar{y} + b_n = 0$ for all n, then $\bar{y}$ is trivially periodic. In this section, we shall look for the existence of nontrivial periodic solutions.

Proposition 3.4.1. *If $A(n)$ is periodic, then the fundamental matrix Φ satisfies the relations*

$$\begin{aligned} \Phi(n+N, N) &= \Phi(n, 0) \\ \Phi(n+N, 0) &= \Phi(n, 0)\Phi(N, 0). \end{aligned} \tag{3.4.1}$$

Proof. The proof follows easily from the definition of Φ (see 3.2.1) and the hypothesis of periodicity of $A(n)$. ∎

Theorem 3.4.1. *If the homogeneous equation* (3.1.2) *has only $y_n = 0$ as periodic solution, then the nonhomogeneous equation* (3.1.1) *has a unique periodic solution of period N and vice versa.*

Proof. If y_n is a periodic solution for (3.1.2) and (3.1.1), it must satisfy respectively

$$y_N = y_0 = \Phi(N, 0)y_0$$

$$y_N = y_0 = \Phi(N, 0)y_0 + \sum_{j=0}^{N-1} \Phi(N, j+1)b_j,$$

from which it follows that y_0 must satisfy

$$By_0 = 0 \tag{3.4.2}$$

$$By_N = \sum_{j=0}^{N-1} \Phi(N, j+1)b_j, \tag{3.4.3}$$

where $B = I - \Phi(N, 0)$.

The solution of each of the two equations (3.4.2) and (3.4.3) will give the initial condition for the periodic solutions of the equations (3.1.2) and (3.1.1).

If (3.4.2) has the unique solution $y_0 = 0$, it follows that $\det B \neq 0$ and this implies that (3.4.3) has a unique nontrivial solution. The converse is proved similarly.

Suppose now that $\det B = 0$ and $N(B)$, the null space of B, has dimension $k < s$. This means that the equation (3.4.2) has k solutions to which will correspond k periodic solutions of (3.1.2). The problem (3.4.3) has solutions if the vector $\sum_{j=0}^{N-1} \Phi(N, j+1)b_j$ is orthogonal to $N(B^T)$. Suppose that $v^{(1)}, v^{(2)}, \ldots, v^{(k)}$ is a base of $N(B^T)$, then we have

$$v^{(i)^T} - v^{(i)^T}\Phi(N, 0) = 0, \qquad i = 1, 2, \ldots, k, \tag{3.4.4}$$

from which

$$v^{(i)^T} = v^{(i)^T}\Phi(N, 0), \qquad i = 1, 2, \ldots, k. \tag{3.4.5}$$

By imposing the orthogonality conditions we get

$$v^{(i)^T} \sum_{j=0}^{N-1} \Phi(N, j+1) b_j = \sum_{j=0}^{N-1} v^{(i)^T} \Phi(N, j+1) b_j = 0. \tag{3.4.6}$$

Let

$$x_j^{(i)^T} = v^{(i)^T} \Phi(N, j+1), \qquad i = 1, 2, \ldots, k, \tag{3.4.7}$$

so that we obtain

$$\sum_{j=0}^{N-1} x_j^{(i)^T} b_j = 0, \qquad i = 1, 2, \ldots, k. \qquad \blacksquare \tag{3.4.8}$$

From this result we get the following theorem.

Theorem 3.4.2. *If the homogeneous equation* (3.1.2) *has k periodic solutions of period N and if the conditions* (3.4.8) *are verified, then the nonhomogeneous equation* (3.1.1) *has periodic solutions of period N.*

Let us now consider the functions defined by (3.4.7) and let x_j, $j \in \mathbf{N}_{n_0}^+$ be one of them. Then,

$$x_{j-1}^T = v^T \Phi(N, j) = v^T \Phi(N, j+1) \Phi(j+1, j).$$

The vector x_j are periodic with period N and satisfy the equation

$$x_{j-1}^T = x_j^T A(j), \tag{3.4.9}$$

which is called the adjoint equation of (3.1.1). The fundamental matrix $\Psi(t, s)$ for such equation satisfies the equation

$$\Psi(t-1, s) = \Psi(t, s) A(t), \qquad \Psi(s, s) = I. \tag{3.4.10}$$

Using (3.4.9), Theorem 3.4.2 can be rephrased as follows.

Theorem 3.4.2′. *If the homogeneous equation* (3.1.2) *has k periodic solutions of period N and if the given vector* $b = (b_0, b_1, \ldots, b_{N-1})^T$ *is orthogonal to the periodic solutions of the adjoint equation, then the nonhomogeneous equation* (3.1.1) *has periodic solutions of period N.*

Theorem 3.4.3. *Suppose* $A(n)$ *and* b_n *periodic with period N. If the nonhomogeneous equation* (3.1.1) *does not have periodic solutions with period N, it cannot have bounded solutions.*

Proof. If (3.1.1) has no period solutions, by Theorem 3.4.1, the equation (3.1.2) has such solutions and of course the conditions (3.4.8) are not verified. Let v^T be a solution of $v^T B = 0$. Then $v^T \sum_{j=0}^{N-1} \Phi(N, j+1) b_j \neq 0$. Consequently, for every solution y_n of (3.1.1), we have

$$v^T y_N = v^T \Phi(N, 0) y_0 + v^T \sum_{j=0}^{N-1} \Phi(N, j+1) b_j$$

$$= v^T y_0 + v^T \sum_{j=0}^{N-1} \Phi(N, j+1) b_j.$$

Moreover, by considering the periodicity of b_j and (3.4.1) we get

$$v^T y_{2N} = v^T y_N + v^T \sum_{j=0}^{N-1} \Phi(N, j+1) b_j$$

$$= v^T y_0 + 2v^T \sum_{j=0}^{N-1} \Phi(N, j+1) b_j,$$

and in general, for $k > 0$,

$$v^T y_{kN} = v^T y_0 + kv^T \sum_{j=0}^{N-1} \Phi(N, j+1) b_j$$

showing that y_n cannot be bounded. ■

The matrix $U \equiv \Phi(N, 0)$ has relevant importance in discussing the stability of periodic solutions. From

$$\Phi(n+N, 0) = \Phi(n+N, N)\Phi(N, 0)$$

and (3.4.1), it follows that:

$$\Phi(n+N, 0) = \Phi(n, 0) U \tag{3.4.10}$$

and in general, for $k > 0$,

$$\Phi(n+kN, 0) = \Phi(n, 0) U^k. \tag{3.4.11}$$

Suppose that ρ is an eigenvalue of U and v the corresponding eigenvector. Then,

$$\Phi(n+N, 0) v = \Phi(n, 0) U v = \rho \Phi(n, 0) v.$$

Letting $\Phi(n, 0) v = v_n$ for $n \geq 0$, we get

$$v_{n+N} = \rho v_n. \tag{3.4.12}$$

This means that the solution of the homogeneous equation having initial value v_n, after one period, is multiplied by ρ. For this reason the eigenvalues of U are usually called multiplicators. The converse is also true. If y_n is a solution such that $y_{n+N} = \rho y_n$, for all n, then in particular is $y_N = \rho y_0$ and that means $U y_0 = \rho y_0$ from which follows that y_0 is an eigenvector of U.

3.5. Boundary Value Problems

The discrete analog of the Sturm–Liouville problem is the following:

$$\Delta(p_{k-1}\Delta y_{k-1}) + (q_k + \lambda r_k)y_k = 0, \tag{3.5.1}$$

$$\alpha_0 y_0 + \alpha_1 y_1 = 0, \quad \alpha_M y_M + \alpha_{M+1} y_{M+1} = 0, \tag{3.5.2}$$

where all the sequences are of real numbers, $r_k > 0$, $\alpha_0 \neq 0$, $\alpha_M \neq 0$ and $0 \leq k \leq M$. The problem can be treated by using arguments very similar to the continuous case. We shall transform the problem into a vector form, and reduce it to a problem of linear algebra. Note that the equation (3.5.1) can be rewritten as

$$p_k y_{k+1} - (p_k + p_{k-1})y_k + p_{k-1}y_{k-1} + (q_k + \lambda r_k)y_k = 0.$$

Let

$$a_k = p_k + p_{k-1} - q_k, \qquad k = 2, \ldots, M-1,$$

$$a_1 = p_1 + p_0 - q_1 + \frac{\alpha_1}{\alpha_0} p_0, \qquad a_M = p_M + p_{M-1} - q_M + \frac{\alpha_M}{\alpha_{M+1}} p_M,$$

$$y = (y_1, y_2, \ldots, y_M)^T, \qquad \boldsymbol{R} = \mathrm{diag}(r_1, r_2, \ldots, r_M),$$

and

$$A = \begin{pmatrix} a_1 & -p_1 & 0 & \ldots & 0 \\ -p_1 & a_2 & -p_2 & & \vdots \\ 0 & & & & 0 \\ \vdots & & & & -p_{M-1} \\ 0 & \ldots & 0 & -p_{M-1} & a_M \end{pmatrix}, \tag{3.5.3}$$

then the problem (3.5.1), (3.5.2) is equivalent to

$$Ay = \lambda R y. \tag{3.5.4}$$

This is a generalized eigenvalue problem for the matrix A. The condition for existence of solutions to this problem is

$$\det(A - \lambda R) = 0, \tag{3.5.5}$$

which is a polynomial equation in λ.

Theorem 3.5.1. *The generalized eigenvalues of* (3.5.4) *are real and distinct.*

Proof. Let $S = \boldsymbol{R}^{-1/2}$. It then follows that the roots of (3.5.5) are roots of

$$\det(SAS - \lambda I) = 0. \tag{3.5.6}$$

Since the matrix SAS is symmetric, it will have real and distinct eigenvalues.

For each eigenvalue λ_i, there is an eigenvector y^i, which is the solution of (3.5.4). By using standard arguments, it can be proved that if y^i and y^j are two eigenvectors associated with two distinct eigenvalues, then

$$(y^i, \boldsymbol{R}y^i) \equiv \sum_{s=1}^{M} r_s y_s^i y_s^j = 0. \quad \blacksquare \tag{3.5.7}$$

Definition 3.5.1. Two vectors u and v such that $(u, \boldsymbol{R}v) = 0$ are called $\boldsymbol{R}$-orthogonal.

Since the Sturm–Liouville problem (3.5.1), (3.5.2) is equivalent to (3.5.4), we have the following result.

Theorem 3.5.2. *Two solutions of the discrete Sturm–Liouville problem corresponding to two distinct eigenvalues are **R**-orthogonal.*

Consider now the more general problem

$$y_{n+1} = A(n)y_n + b_n, \tag{3.5.8}$$

where y_n, $b_n \in \boldsymbol{R}^s$ and $A(n)$ is an $s \times s$ matrix. Assume the boundary conditions are given by

$$\sum_{i=0}^{N} L_i y_{n_i} = w, \tag{3.5.9}$$

where $n_i \in N_0^+$, $n_i < n_{i+1}$, $n_0 = 0$, w is a given vector in $\boldsymbol{R}^s$ and L_i are given $s \times s$ matrices. Let $\Phi(n, j)$ be the fundamental matrix for the homogeneous problem

$$y_{n+1} = A(n)y_n, \tag{3.5.10}$$

such that $\Phi(0, 0) = I$. The solutions of (3.5.8) are given

$$y_n = \Phi(n, 0)y_0 + \sum_{j=0}^{n-1} \Phi(n, j+1)b_j, \tag{3.5.11}$$

where y_0 is the unknown initial condition. The conditions (3.5.9) will be satisfied if

$$\sum_{i=0}^{N} L_i y_{n_i} = \sum_{i=0}^{N} L_i \Phi(n_i, 0)y_0 + \sum_{i=0}^{N} L_i \sum_{j=0}^{n_i-1} \Phi(n_i, j+1)b_j = w,$$

which can be written as

$$\sum_{i=0}^{N} L_i \Phi(n_o, 0) y_0 + \sum_{i=0}^{N} L_i \sum_{j=0}^{n_N - 1} \phi(n_i, j+1) T(j+1, n_i) b_j = w, \tag{3.5.12}$$

where the step matrix $T(j, n)$ is defined by

$$T(j, n) = \begin{cases} I & \text{for } j \leq n, \\ 0 & \text{for } j > n. \end{cases}$$

By introducing the matrix $Q = \sum_{i=0}^{N} L_i \Phi(n_i, 0)$, the previous formula becomes

$$Qy_0 = w - \sum_{i=0}^{N} \sum_{j=0}^{n_N - 1} L_i \Phi(n_i, j+1) T(j+1, n_i) b_j. \tag{3.5.13}$$

Theorem 3.5.3. *If the matrix Q is nonsingular, then the problem* (3.5.8) *with boundary conditions* (3.5.9) *has only one solution given by*

$$y_n = \Phi(n, 0) Q^{-1} w + \sum_{s=0}^{n_N - 1} G(n, s) b_s, \tag{3.5.14}$$

where the matrices $G(n, j)$ are defined by:

$$\begin{aligned} G(n, j) &= \Phi(n, j+1) T(j+1, n) \\ &\quad - \Phi(n, 0) Q^{-1} \sum_{i=0}^{N} L_i \Phi(n_i, j+1) T(j+1, n_i). \end{aligned} \tag{3.5.15}$$

Proof. Since Q is nonsingular, from (3.5.13) we see that (3.5.11) solves the problem if the initial condition is given by

$$y_0 = Q^{-1} w - Q^{-1} \sum_{i=0}^{N} \sum_{j=0}^{n_N - 1} L_i \Phi(n_i, j+1) T(j+1, n_i) b_j. \tag{3.5.16}$$

By substituting in (3.5.11) one has

$$\begin{aligned} y_n &= \Phi(n, 0) Q^{-1} w - \Phi(n, 0) Q^{-1} \sum_{i=0}^{N} \sum_{j=0}^{n_N - 1} L_i \Phi(n_i, j+1) T(j+1, n_i) b_j \\ &\quad + \sum_{j=0}^{n_N - 1} \Phi(n, j+1) T(j+1, n) b_j \\ &= \Phi(n, 0) Q^{-1} w + \sum_{j=0}^{n_N - 1} \Bigg[\Phi(n, j+1) T(j+1, n) \\ &\quad - \Phi(n, 0) Q^{-1} \sum_{i=0}^{N} L_i \Phi(n_i, j+1) T(j+1, n_i) \Bigg] b_j, \end{aligned}$$

from which by using the definition (3.5.14) of $G(n, j)$, the conclusion follows ■

The matrix $G(n, i)$ is called Green's matrix and it has some interesting properties. For example,

(1) for fixed j, the function $G(n, j)$ satisfies the boundary conditions $\sum_{i=0}^{N} L_i G(n_i, j) = 0$,
(2) for fixed j and $n \neq j$, the function $G(n, j)$ satisfies the autonomous equation $G(n+1, j) = A(n)G(n, j)$, and
(3) for $n = j$, one has $G(j+1, j) = A(j)G(j, j) + I$.

The proofs of the above statements are left as exercizes. (See problems 3.19, 3.20).

If the matrix Q is singular then the equation (3.5.15) can have either an infinite number of solutions or no solution. Suppose, for simplicity, we indicate by b the righthand side of (3.5.13), then the problem is reduced to establishing the existence of solutions for the equation

$$Qy_0 = b. \tag{3.5.17}$$

Let $\boldsymbol{R}(Q)$ and $\boldsymbol{N}(Q)$ be respectively the range and the null space of Q. Then (3.5.17) will have solutions if $b \in \boldsymbol{R}(Q)$. In this case if c is any vector in $\boldsymbol{N}(Q)$ and $\bar{y}_0$ any solution of (3.5.17), the vector $c + \bar{y}_0$ will also be a solution. Otherwise if $b \notin \boldsymbol{R}(Q)$, the problem will not have solutions.

In the first case ($b \in \boldsymbol{R}(Q)$), a solution can be obtained by introducing the generalized inverse of Q, defined as follows. Let $r = \text{rank } Q$. The generalized inverse Q^I of Q is the only matrix satisfying the relations

$$QQ^I = P, \qquad Q^I Q = P_1, \tag{3.5.18}$$

$$QQ^I Q = Q, \quad Q^I Q Q^I = Q^I, \tag{3.5.19}$$

where P and P_1 are the projections on $\boldsymbol{R}(Q)$ and $\boldsymbol{R}(Q^*)$ (Q^* is the conjugate transpose of Q) respectively. It is well known that if F is an $s \times s$ matrix whose columns span $\boldsymbol{R}(Q)$, then P is given by

$$F(F^*F)^{-1}F^*. \tag{3.5.20}$$

By using Q^I the solution $\bar{y}_0$ of (3.5.17) when $b \in \boldsymbol{R}(Q)$ is given by

$$\bar{y}_0 = Q^I b. \tag{3.5.21}$$

In fact, we have $Q\bar{y}_0 = QQ^I b = Pb = b$. A solution $\bar{y}_n$ of the boundary value problem can now be given in a form similar to (3.5.14) and (3.5.15) with Q^{-1} replaced by Q^I. This solution, as we have already seen, is not unique.

In fact if $c \in N(Q)$, $y_k = \Phi(n, 0)c + \bar{y}_n$ will also be a solution satisfying the boundary conditions since

$$\sum_{i=0}^{N} L_i y_{n_i} = \sum_{i=0}^{N} \Phi(n_i, 0)c + \sum_{i=0}^{n} L_i \bar{y}_{n^i} = Qc + w = w.$$

When $b \notin R(Q)$, the relation (3.5.21) has the meaning of least square solution, because $\bar{y}_0$ minimizes the quantity $\|Qy_0 - b\|_2$ and the sequence $\bar{y}_n$ defined consequently may serve as an approximate solution.

3.6. Problems

3.1. Show that the vectors $f_1(n) = (1, n)^T$ and $f_2(n) = (n, n^2)^T$ are linearly independent even if the matrix $K(n) = (f_1(n), f_2(n))$ has determinant always zero. Can the two vectors be solutions of (3.1.2)?

3.2. Prove that if $A(n)$ is nonsingular for $n \geq n_0$, then $\Phi^{-1}(n, s)$ exists for $n, s \geq n_0$. (Hint: write $\Phi(n, s) = \prod_{i=s}^{n-1} A(i)$ and invert.)

3.3. Show that if $K(n)$ satisfies (3.1.5), with $K(n) \neq 0$ for all n, its column form a set of linearly independent solutions of (3.1.2).

3.4. Prove that if A is a constant matrix, $\Phi(n, n_0) = \Phi(n - n_0)$ either directly by the explicit expression of $\Phi(n, n_0)$ or as a solution of (3.1.5).

3.5. Prove the relations (3.3.14) and (3.3.15).

3.6. Prove proposition 3.3.1 directly by using the expression (3.3.11).

3.7. Verify that (3.3.16) satisfies equation (3.3.1).

3.8. Verify that $y_n = \sum_{s=-\infty}^{-1} A^s b_{n-s-1}$ is a solution of $y_{n+1} = Ay_n + b_n$. When does this solution have a meaning?

3.9. Verify that $H(n)$ is solution of $\sum_{i=0}^{k} p_i y_{n+k-i} = 0$.

3.10. Supposing that $\sum_{j=0}^{\infty} H(n-j)g_j$ has a meaning, show that it is the solution of the autonomous linear scalar difference equation. When does it have a meaning?

3.11. As in the previous exercise with the function $y_n = \sum_{y=-\infty}^{+\infty} H(n-j)g_j$, where $H(n)$ is the solution of $\sum_{i=0}^{k} p_i H(n+k-i) = \delta_n^{-1}$, find the conditions on the roots of $p(z)$ in order to have y_n as the only bounded solution.

3.12. Consider the second order difference equation $y_{n+2} + p_1 y_{n+1} + p_2 y_n = g_n$. Construct the function $H(n)$ of the previous exercise.

3.13. Find the function $H(n-j)$ and the solution of $y_{n+2} - 2(\cos t)y_{n+1} + y_n = g_n$, $y_0 = y_1 = 0$.

3.14. By using the onesided Green's function H defined by (3.3.17) find the inverse of the matrix:

$$C_N = \begin{pmatrix} \alpha & \beta & & & \\ \gamma & \cdot & \cdot & & \\ & \cdot & \cdot & \cdot & 0 \\ & 0 & & & \beta \\ & & & \gamma & \alpha \end{pmatrix}_{N\times N}$$

When does the inverse exist for $N \to \infty$ (infinite matrix)?

3.15. Deduce the adjoint equation in the scalar case as defined in section 2.1 from the Definition 3.4.9.

3.16. Show that $\Delta^2 y_n + 4\sin^2 \frac{\pi}{2M} y_n = 0$, $y_0 = 0$, $y_M = \varepsilon$ $(\neq 0)$, has no solution. It is interesting to notice that the continuous analog of this problem, namely, $y'' + 4y\sin^2 \frac{\pi}{2M} = 0$, $y_0 = 0$, $y(M) = \varepsilon$, has solution $y(t) = \dfrac{\sin\left(2t\sin\frac{\pi}{2M}\right)}{\sin\left(2M\sin\frac{\pi}{2M}\right)}\varepsilon$.

3.17. Find the solution of $\Delta^2 y_n + \frac{\pi^2}{M^2} y_n = 0$, $y(0) = y(M) = 0$, for $M > 2$.

3.18. Find the eigenvalues of $\Delta^2 y_n + \lambda y_n = 0$, $y_0 = 0$, $y_{M+1} = 0$. Show that the problem is equivalent to finding the eigenvalues of the matrix:

$$A = \begin{pmatrix} 2 & -1 & 0 & & \cdots & 0 \\ -1 & 2 & -1 & 0 & \cdots & 0 \\ 0 & & & & & \vdots \\ \vdots & & & & & -1 \\ 0 & & \cdots & 0 & -1 & 2 \end{pmatrix}_{M\times M}.$$

3.19. Show that for fixed j and $n \neq j$, $G(n, j)$ satisfies the homogeneous equation $G(n+1, j) = AG(n, j)$.

3.20. Show that for $j+1$ and j in the same interval $[n_{i-1}, n_i - 1]$ one has $G(j+1, j) = A(j)G(j, j) + I$.

3.21. Show that for fixed j, $G(n, j)$ satisfies the boundary conditions $\sum_{i=0}^{N} L_i G(n_i, j) = 0$.

3.22. Verify that (3.5.14) satisfies the Equation (3.5.8) and Conditions (3.5.9).

3.7. Notes

Most of the material of Sections 3.1, 3.2 and 3.3 appears in several different books. The major references are Miller [115] and Hahn [65]. The Poincaré and Perron's theorems can be found in Gelfond [57, 58]. The periodic solutions are treated in Halanay [69], Halanay and Wexler [68] and Corduneanu [31]. The boundary value problems are treated in Fort [49], Mattheij [104, 105], Agarwal [3, 4] and Hildebrand [78]. Theorem 3.5.2 has been taken from Agarwal [3].

CHAPTER 4 Stability Theory

4.0. Introduction

In Section 4.1, we define various stability notions and give some simple examples. Sections 4.2 to 4.4 are devoted to the theory of stability of linear difference equations. Using norm as a candidate for measure and comparison principle, we develop general results on stability in Section 4.5. Nonlinear variation of constants formula is obtained in Section 4.6 and stability by first approximation is treated in Section 4.7. Sections 4.8 and 4.9 investigate stability theory in terms of Lyapunov functions and comparison principle, and offer several direct theorems of importance. A discussion of converse theorems is the content of Section 4.10. Section 4.11 deals with the concepts of total and practical stabilities that are important in applications to numerical analysis. Several problems are included in Section 4.12 that complement the theory that is presented in the chapter.

4.1. Stability Notions

Let us denote by $B(y, \delta)$ the open ball having center at y and radius δ. If $y = 0$, we shall use the notation B_δ. Let $y_0 \in B_a$ and $f: \boldsymbol{R}^s \to \boldsymbol{R}^s$ be a bounded function in B_a. The solution of the difference equation

$$y_{n+1} = f(n, y_n), \qquad y_{n_0} = y_0, \tag{4.1.1}$$

will remain in B_a for all $n \geq n_0$ such that $f(n, y_n) \in B_a$. Let $I_{n_0} \subset \boldsymbol{N}^+_{n_0}$ be the set of all such indicies.

Definition 4.1.1. The points $\bar{y} \in \mathbf{R}^s$, which satisfy the algebraic equation

$$f(n, \bar{y}) = \bar{y}$$

for all n, are called fixed points (or critical or limit points) of (4.1.1).

We shall suppose, for simplicity, that there is a fixed point at the origin. If the fixed point is not at the origin, changing coordinates $z_n = y_n - \bar{y}$, equation (4.1.1) is transformed into

$$z_{n+1} = f(n, z_n + \bar{y}) - f(n, \bar{y}) \equiv F(n, z_n),$$

which has the fixed point at the origin.

Definition 4.1.2. The solution $y = 0$ of (4.1.1) is said to be

(i) stable if given $\varepsilon > 0$, there is a $\delta(\varepsilon, n_0)$ such that for any $y_0 \in B_\delta$ the solution $y_n \in B_\varepsilon$.

(ii) Uniformly stable if it is stable and δ can be chosen independent on n_0.

(iii) Attractive if there is $\delta(n_0) > 0$ such that for $y_0 \in B_\delta$ one has $\lim_{n\to\infty} y_n = 0$.

(iv) Uniformly attractive if it is attractive and δ can be chosen independent of n_0.

(v) Asymptotically stable if it is stable and attractive.

(vi) Uniformly asymptotically stable if it is uniformly stable and uniformly attractive.

(vii) Globally attractive if it is attractive for all starting points $y_0 \in \mathbf{R}^s$.

(viii) Globally asymptotically stable if it is asymptotically stable for all $y_0 \in \mathbf{R}^s$.

(ix) Exponentially stable if there exists $\delta > 0$, $a > 0$, $\eta \in (0, 1)$ such that if $y_0 \in B_\delta$, then

$$\|y_n\| \leq a\|y_0\|\eta^{n-n_0}.$$

(x) *lp*-stable if it is stable and moreover for some $p > 0$,

$$\sum_{j=n_0}^{\infty} \|y(j, n_0, y_0)\|^p < \infty.$$

(xi) Uniformly *lp*-stable if the previous summation converges uniformly with respect to n_0.

Example. Consider the difference equation

$$x_{n+1} = a_n x_n, \qquad x_{n_0} = x_0, \tag{4.1.2}$$

where a_i are real numbers. The solution of (4.1.2) is

$$x_n = x_0 \prod_{i=n_0}^{n-1} a_i. \tag{4.1.3}$$

We then have the following cases:

$$\text{(a) If } \left| \prod_{i=n_0}^{n-1} a_i \right| < M(n_0), \tag{4.1.4}$$

then

$$|x_n| < |x_0| M(n_0)$$

and it suffices to take

$$\delta(\varepsilon, n_0) = \frac{\varepsilon}{M(n_0)}$$

to get stability.

$$\text{(b) If } \left| \prod_{i=n_0}^{n-1} a_i \right| \leq M, \tag{4.1.5}$$

then it suffices to take $\delta = \varepsilon / M$ to get uniform stability. This condition is satisfied for example if $a_i = \cos i$.

$$\text{(c) If } \lim_{n \to \infty} \left| \prod_{i=n_0}^{n-1} a_i \right| = 0, \tag{4.1.6}$$

then, as this is a particular case of (4.1.5), the stability follows and moreover,

$$\lim_{n \to \infty} |x_n| = |x_0| \lim_{n \to \infty} \left| \prod_{i=n_0}^{n-1} a_i \right| = 0,$$

obtaining the asymptotic stability of the zero solution. This condition is not satisfied when $a_i = \cos i$ as in the previous case, showing that uniform stability and asymptotic stability are two different concepts.

$$\text{(d) If } \left| \prod_{i=n_0}^{n-1} a_i \right| < a\eta^{n-n_0}, \tag{4.1.7}$$

where $a > 0$ and $0 < \eta < 1$, then $x = 0$ is exponentially stable.

There is a sort of hierarchy among all these different kinds of stabilities. For example, uniform asymptotic stability implies asymptotic stability and uniform stability implies stability. Furthermore, asymptotic stability implies stability.

Equation (4.1.1) is said to be autonomous if f does not depend explicitly on n. For autonomous equations uniform stability concepts coincide with the stability ones. This can be seen by observing that if $y(n, n_0, y_0)$ satisfies the autonomous equation $y(n+1, n_0, y_0) = f(y(n, n_0, y_0))$, then $y(n - n_0, 0, y_0)$ satisfies the same equation. Since the two solutions assume the same value for $n = n_0$, they must coincide for $n \in I_{n_0}$. This means that for autonomous equations we can always fix $n_0 = 0$ and if there is stability for $n_0 = 0$, then it is true for all n_0, which means that stability is uniform.

Attractivity is a different concept than stability, as the following example demonstrates.

Consider the system

$$
\begin{aligned}
x_{n+1} &= x_n + \frac{x_n^2(y_n - x_n) + y_n^5}{r_n^2 + r_n^6} \equiv x_n + g_1(x_n, y_n), \\
y_{n+1} &= y_n + \frac{y_n^2(y_n - 2x_n)}{r_n^2 + r_n^6} \equiv y_n + g_2(x_n, y_n),
\end{aligned} \tag{4.1.8}
$$

where $r_n^2 = x_n^2 + y_n^2$. The origin of (4.1.10) is globally attractive but unstable (see problem 4.3). Of course lp-stability implies asymptotic stability, because the convergence of the series $\Sigma\|y(j, n_0, y_0)\|^p$ will imply that $y(j, n_0, y_0)$ tends to zero. The converse is not true. Exponential stability, however, is sufficient for lp-stability as the next result shows.

Theorem 4.1.1. *If the solution $y = 0$ is exponentially stable, then it is also lp-stable.*

Proof. By definition we have:

$$\|y_n\| \leq a\|y_0\|\eta^{n-n_0}$$

with $0 < \eta < 1$, and $a > 0$. Therefore $\sum_{j=n_0}^{\infty}\|y_j\|^p \leq a^p\|y_0\|^p \sum_{j=n_0}^{\infty}(\eta^p)^{j-n_0} \leq$

$$a^p\|y_0\|^p \frac{1}{1-\eta^p}. \qquad \blacksquare$$

4.2. *The Linear Case*

We shall discuss two results that characterize both uniform stability and uniform asymptotic stability in terms of the fundamental matrix. Consider

$$y_{n+1} = A(n)y_n, \qquad y_{n_0} = y_0, \tag{4.2.1}$$

where $A(n)$ is an $s \times s$ matrix.

Theorem 4.2.1. *Let $\Phi(n, n_0)$ be the fundamental matrix of* (4.2.1). *Then the solution $y = 0$ is uniformly stable if there exists an $M > 0$ such that*

$$\|\Phi(n, n_0)\| < M, \qquad \text{for } n \geq n_0. \tag{4.2.2}$$

Proof. The sufficiency follows from the fact that $y_n = \Phi(n, n_0)y_0$. For, we have

$$\|y_n\| \leq \|\Phi(n, n_0)\| \, \|y_0\| \leq M\|y_0\|,$$

and hence $\|y_n\| < \varepsilon$ if $\|y_0\| < \varepsilon M^{-1}$.

To prove necessity, if there is uniform stability, then

$$\|\Phi(n, n_0)y_0\| < 1$$

for $\|y_0\| < \delta$. Taking $x_0 = \dfrac{y_0}{\|y_0\|}$ the previous formula shows that

$$\sup_{\|x_0\|=1} \|\Phi(n, n_0)x_0\| \tag{4.2.3}$$

is bounded. But (4.2.3) is just the definition of the norm of $\Phi(n, n_0)$ (see Appendix A). ■

Theorem 4.2.2. *The solution $y = 0$ of* (4.2.1) *is uniformly asymptotically stable if there exists two positive numbers a, η with $\eta < 1$ such that*

$$\|\Phi(n, n_0)\| \leq a\eta^{n-n_0}.$$

Proof. The proof of the sufficiency is as simple as before. The necessity follows by considering that if there is uniform asymptotic stability, then fixing $\varepsilon > 0$ there exists $\delta > 0$, $\boldsymbol{N}(\varepsilon) > 0$ such that for $y_0 \in B_\delta$,

$$\|\Phi(n, n_0)y_0\| < \varepsilon$$

for $n \geq n_0 + \boldsymbol{N}(\varepsilon)$. As before, it is easy to see that

$$\|\Phi(n, n_0)\| < \eta$$

for $n \geq n_0 + \boldsymbol{N}(\varepsilon)$, where this time η can be chosen arbitrarily small. Moreover, because the uniform asymptotic stability implies the uniform stability, we obtain $\|\Phi(n, n_0)\|$ is bounded by a positive number a for all $n \geq n_0$. We then have for $n \in [n_0 + m\boldsymbol{N}(\varepsilon), n_0 + (m+1)\boldsymbol{N}(\varepsilon)]$,

$$\|\Phi(n, n_0)\| \leq \|\Phi(n, n_0 + m\boldsymbol{N}(\varepsilon))\| \, \|\Phi(n_0 + m\boldsymbol{N}(\varepsilon), n_0 + (m-1)\boldsymbol{N}(\varepsilon)\| \ldots$$

$$\ldots \|\Phi(n_0 + \boldsymbol{N}(\varepsilon), n_0)\| < a\eta^m = \alpha\eta^{-1}\eta\frac{M+1}{\boldsymbol{N}(\varepsilon)}\boldsymbol{N}(\varepsilon) = \alpha_1\eta_1^{\eta-n_0}$$

with $\eta_1 = \eta^{\frac{1}{N(\varepsilon)}} < 1$, $\alpha_1 = \alpha\eta^{-1}$ and this proves the theorem. ■

As a result of this theorem we see that for linear systems, uniform asymptotic stability is equivalent to exponential asymptotic stability.

4.3. Autonomous Linear Systems

In this section we shall be concerned with linear autonomous equations because of their importance in applications. Of course, the results of theorems 4.2.1 and 4.2.2 hold true, but we can give more explicit results in this special case.

The solution of the homogeneous autonomous equation

$$y_{n+1} = Ay_n, \qquad y_{n_0} = y_0, \tag{4.3.1}$$

is given by

$$y_n = A^{n-n_0}y_0. \tag{4.3.2}$$

From the matrix theory (see Appendix A), we know that

$$A^q = \sum_{k=1}^{r} \sum_{i=0}^{m_k-1} \binom{q}{i} \lambda_k^{q-i}(A - \lambda_k I)^i Z_{k_1}, \tag{4.3.3}$$

where r is the number of distinct eigenvalues of A, m_k is the multiplicity of λ_k and Z_{k_1} are component matrices of A, which are independent of λ_k. From (4.3.3) it follows that if the eigenvalues of A are in the unit disk, then $\lim_{n\to\infty} A^{n-n_0} = 0$ and vice versa. This leads to the following result.

Theorem 4.3.1. *The solution $y = 0$ of (4.3.1) is asymptotically stable iff the eigenvalues of the matrix A are inside the unit disk.*

We recall that an eigenvalue λ_k is said to be semisimple if $(A - \lambda_k I)Z_{k_1} = 0$ (see Appendix A).

Theorem 4.3.2. *The solution $y = 0$ of (4.3.1) is stable iff the eigenvalues of A have modulus less or equal to one, and those of modulus one are semisimple.*

Proof. From (4.3.3) it is easy to see that for semisimple eigenvalues the term containing q is $\binom{q}{0} = 1$, which does not grow for $q \to \infty$. ■

If the matrix A is a companion matrix, then it is known (see Appendix A) that there are no semisimple eigenvalues that are not simple. In this case Theorem 4.3.2 assumes the following form, which is very useful in numerical analysis.

Theorem 4.3.3. *If A is a companion matrix, the solution $y = 0$ is stable iff the eigenvalues of A have modulus less or equal to 1 and those of modulus 1 are simple.*

Example. The trivial equation

$$y_{n+1} = Iy_n, \tag{4.3.4}$$

where I is the $s \times s$ unit matrix is an example where the matrix has a multiple eigenvalue that is 1, but it is semisimple and the zero solution is stable.

Let us now consider the nonhomogeneous equation

$$y_{n+1} = Ay_n + b, \tag{4.3.5}$$

where A is an $s \times s$ nonnegative matrix and b a nonnegative vector. The critical point $\bar{y}$ is given by the solution of the equation

$$\bar{y} = A\bar{y} + b. \tag{4.3.6}$$

For such difference equations, there is a relation between the existence of nonnegative solutions of (4.3.6) and the stability behavior. For the notion used in the next two theorems see Appendix A.

Theorem 4.3.4. *If $A \geq 0$, $b \geq 0$ and $\rho(A) < 1$, where $\rho(A)$ is the spectral radius of A, then* (4.3.6) *has a nonnegative solution.*

Proof. Since $\rho(A) < 1$, $(I - A)^{-1}$ exists and is given by $(I - A)^{-1} = \sum_{i=0}^{\infty} A^i$ from which we obtain $\bar{y} = \sum_{i=0}^{\infty} A^i b$, which is nonnegative. By assumption on $\rho(A)$, we also see that $\bar{y}$ is asymptotically stable ■

Under stronger assumptions on b, there is a converse of the above theorem.

Theorem 4.3.5. *Suppose that b is positive and $A \geq 0$, then, if* (4.3.6) *has a positive solution $\bar{y}$, we have $\rho(A) < 1$.*

Proof. By the Perron–Frobenius Theorem the matrix A^T has a real eigenvalue equal to $\rho(A)$ to which corresponds a nonnegative eigenvector u_0 such that $A^T u_0 = \rho(A)u_0$. Multiplying the transpose of the relation (4.3.6) by u_0, we get $[1 - \rho(A)]\bar{y}^T u_0 = b^T u_0$ from which, since both $\bar{y}^T u_0$ and $b^T u_0$ are positive, one gets $\rho(A) < 1$. ■

The foregoing results have important applications in the study of iterative methods for linear systems of equations.

4.4. *Linear Equations with Periodic Coefficients*

The results obtained in the previous section cannot be extended to nonautonomous equations, as the following example shows. Consider the equation

$$y_{n+1} = A(n)y_n, \tag{4.4.1}$$

where

$$A(n) = \frac{1}{8}\begin{pmatrix} 0 & 9 + (-1)^n 7 \\ 9 - (-1)^n 7 & 0 \end{pmatrix}. \tag{4.4.2}$$

The eigenvalues of $A(n)$ are $\pm 2^{-1/2}$ for all n, and they are inside the unit disk, but this is not enough to ensure even the stability of the null solution. With $n_0 = 0$, the fundamental matrix is

$$\Phi(n, 0) = \begin{pmatrix} 2^{-2n} & 0 \\ 0 & 2^n \end{pmatrix}, \tag{4.4.3}$$

if n is even, and

$$\Phi(n, 0) = \begin{pmatrix} 0 & 2^n \\ 2^{-2n} & 0 \end{pmatrix}, \tag{4.4.4}$$

if n is odd. In any case this is a solution that will grow exponentially away from the origin. Consequently there must be an additional condition on $A(n)$ in order to get stability.

There is an intermediate case, however, that we can treat, namely, linear equations with periodic matrix $A(n)$. The equation (3.4.11) shows that the central role played by the matrix $U \equiv \Phi(N, 0)$ in such cases. Any solution of the equation will have the form

$$y_{n+jN} = \Phi(n, 0)\, U^j y_0 \tag{4.4.5}$$

for $0 \leq n \leq N$. The behavior of the solution will then be dictated by the behavior of $U^j y_0$.

This leads to the following theorem, which is analogous to Theorem 4.2.1.

Theorem 4.4.1. *The zero solution of the equation $y_{n+1} = A(n) y_n$, where $A(n)$ is periodic of period N, is asymptotically stable if the eigenvalues of the matrix U are inside the unit disk. When some semisimple eigenvalues are on the boundary of the unit disk, then the solution is stable. In other cases there is instability.*

The similarity of the results in the two cases of autonomous and periodic equations suggests a more intimate connection between them. This is really the case as the following theorem shows.

Theorem 4.4.2. *If $A(i)$ is, for all i nonsingular and periodic, then it is possible to transform the periodic system into an autonomous one.*

Proof. By hypothesis, the matrix

$$U \equiv \Phi(N, 0) = \prod_{i=0}^{N-1} A(i) \tag{4.4.6}$$

is nonsingular. It is possible to define (see Appendix A) the matrix C such that

$$C = U^{1/N}. \tag{4.4.7}$$

The matrix $P(n) = \Phi(n, 0)C^{-n}$ is periodic because $P(n+N) = \Phi(n+N, 0)C^{-n}C^{-N} = \Phi(n, 0)\Phi(N, 0)C^{-n}C^{-N} = P(n)$. Using this matrix to define the new variable

$$x_n = P^{-1}(n)y_n, \tag{4.4.8}$$

we have $P(n+1)x_{n+1} = A(n)P(n)x_n$ from which we get $x_{n+1} = P_{n+1}^{-1}A(n)P(n)x_n$. Simple manipulations reduce this to

$$x_{n+1} = Cx_n, \tag{4.4.9}$$

which proves the theorem. ■

The solutions having as initial values the eigenvectors of C (or U) have the property

$$x_n = \mu^n v,$$

where v is an eigenvector and μ the corresponding eigenvalue. But $\mu = \rho^{1/N}$, where ρ is the eigenvalue of U. The corresponding solution of the original equation is

$$y_n = \Phi(n, 0)v, \tag{4.4.10}$$

and this agrees with what was stated in Chapter 3.

The solutions in the form of (4.4.10) are said to be Floquet solutions in analogy with the terminology used in the continuous case.

4.5. *Use of the Comparison Principle*

Most of the results on the qualitative behavior of the solutions of difference equations can be obtained using the comparison theorems stated in Section 1.8. This theory is parallel to the corresponding theory of differential equations.

Theorem 4.5.1. *Let $g(n, u)$ be a nonnegative function nondecreasing in u. Suppose that*
(1) $f: N_{n_0}^+ \times B_\rho \to B_\rho, \qquad \rho > 0,$
(2) $f(n, 0) = 0, \qquad g(n, 0) = 0,$ *and*
(3) $\|f(n, y)\| \leq g(n, \|y\|)$.
Then the stability of the trivial solution of the equation

$$u_{n+1} = g(n, u_n) \tag{4.5.1}$$

implies the stability of the trivial solution of (4.1.1).

Proof. From (4.1.1) we have

$$\|y_{n+1}\| \leq \|f(n, y_n)\| \leq g(n, \|y_n\|),$$

and hence the comparison equation is (4.5.1). Theorem 1.6.1 can be applied, provided that $\|y_0\| \leq u_0$, to get $\|y_n\| \leq u_n$ for $n \geq n_0$. Suppose now that the zero solution of (4.5.1) is stable. Then for $\varepsilon > 0$, there exists a $\delta(\varepsilon, n_0)$ such that for $|u_0| < \delta$ we have $|u_n| < \varepsilon$. This means the stability of the trivial solution of (4.1.1).

Since $g \geq 0$, the trivial solution of the comparison equation (4.5.1) may not be, for example, asymptotically stable and therefore we cannot conclude from Theorem 4.5.1 that the trivial solution of (4.1.1) is also asymptotically stable. This is due to the strong assumption (3), which can be replaced by the following condition

(4) $$\|f(n, y)\| \leq \|y\| + w(n, \|y\|),$$

where $g(n, u) = u + w(n, u)$ is nondecreasing in u. Noting that w in (4) need not be positive and hence the trivial solution of $u_{n+1} = g(n, u_n)$ can have different stability properties, we can conclude from Theorem 4.5.1 in case (3) is replaced by (4), that the stability properties of the trivial solution of (4.5.1) imply the corresponding stability properties of the trivial solution of (4.1.1). This version of Theorem 4.5.1 is more useful and we shall denote it by 4.5.1*. Of course, the proof needs minor modifications. ∎

From Theorem 4.5.1 and 4.5.1*, we can easily obtain several important variants.

Theorem 4.5.2. *Let $\Phi(n, n_0)$ be the fundamental matrix of the linear equation*

$$x_{n+1} = A(n)x_n. \tag{4.5.2}$$

Let $F: N_{n_0}^+ \times R^s \to R^s$, $F(n, 0) = 0$ and

$$\|\Phi^{-1}(n+1, n_0)F(n, \Phi(n, n_0)y_n)\| \leq g(n, \|y_n\|), \tag{4.5.3}$$

where the function $g(n, u)$ is nondecreasing in u. Assume that the solutions u_n of

$$u_{n+1} = u_n + g(n, u_n) \tag{4.5.4}$$

are bounded for $n \geq n_0$. Then the stability properties of the linear equation (4.5.2) *imply the corresponding stability properties of the null solution of*

$$x_{n+1} = A(n)x_n + F(n, x_n). \tag{4.5.5}$$

Proof. The linear transformation $x_n = \Phi(n, n_0)y_n$ reduces (4.5.5) to

$$y_{n+1} = y_n + \Phi^{-1}(n+1, n_0)F(n, \Phi(n, n_0)y_n).$$

We then have

$$\|y_{n+1}\| \leq \|y_n\| + g(n, \|y_n\|). \tag{4.5.6}$$

If $\|y_0\| \leq u_0$ we obtain $\|y_n\| \leq u_n$, where u_n is the solution of $u_{n+1} = u_n + g(n, u_n)$. It then follows that $\|x_n\| \leq \Phi(n, n_0)\| \, \|y_n\| \leq \|\Phi(n, n_0)\| u_n$.

If the solution of the linear system is, for example, uniformly asymptotically stable, then from Theorem 4.2.2 we see that $\|\Phi(n, n_0)\| \leq \alpha\eta^{n-n_0}$ for some suitable $\alpha > 0$ and $0 < \eta < 1$. Then

$$\|x_n\| \leq \alpha\eta^{n-n_0}u_n$$

and this shows that the solution $x = 0$ is uniformly asymptotically stable because u_n is bounded. The proof of other cases is similar. ■

We shall merely state another important variant of Theorem 4.5.1* which is widely used in numerical analysis.

Theorem 4.5.3. *Given the difference equation*

$$y_{n+1} = y_n + hA(n)y_n + f(n, y_n), \tag{4.5.7}$$

where h is a positive constant, suppose that

(1) $f(n, 0) = 0$ *for* $n \geq n_0$, *and*
(2) $\|f(n, y)\| \leq g(n, \|y\|)$ *with* $g(n, u)$ *nondecreasing in u,* $g(n, 0) = 0$.

Then the stability properties of the null solution of

$$u_{n+1} = \|I + hA(n)\|u_n + g(n, u_n) \tag{4.5.8}$$

imply the corresponding stability properties of the null solution of (4.5.7).

In this form the foregoing theorem is used in the estimation of the growth of errors in numerical methods for differential equations.

Instead of (4.5.8), one uses usually the comparison equation

$$u_{n+1} = (I + h\|A\|)u_n + g(n, u_n), \tag{4.5.9}$$

which is less useful because $(1+h\|A\|)>1$. The form (4.5.8) is more interesting because when the eigenvalues of A have all negative real parts, $\|I+hA\|$ can be less than 1. This will happen, for example, if the logarithmic norm:

$$\mu(A(n)) = \lim_{h\to 0} \frac{\|I+hA(n)\|-1}{h}$$

is less than zero. From the definition it follows that

$$\|I+hA\| = 1 + h\mu(A(n)) + O(h^2A(n)).$$

Letting

$$\hat{\mu}(A) = h[\mu(A_n) + O(h^2A(n))],$$

the comparison equation becomes

$$u_{n+1} = (1+\hat{\mu})u_n + g(n, u_n). \tag{4.5.10}$$

The next theorem requires essentially a condition on the variation of f, but it does not require the a-priori knowledge of the existence of the critical point.

Let us consider

$$x_{n+1} = f(x_n). \tag{4.5.11}$$

Theorem 4.5.4. *Suppose $f: D \subset \mathbf{R}^s \to \mathbf{R}^s$ is continuous and g is a positive function nondecreasing with respect to its arguments, defined on $J_1 \times J_2 \times J_3$ where J_i are subsets of $\mathbf{R}^+$ containing the origin. Further suppose that for $x_0 \in D$, the sequence x_n is contained in D,*

$$\|x_{n+2} - x_{n+1}\| \leq g(\|x_{n+1} - x_n\|, \|x_{n+1} - x_0\|, \|x_n - x_0\|) \tag{4.5.12}$$

and the comparison equation

$$u_{n+1} = g\left(u_n, \sum_{j=0}^{n} u_j, \sum_{j=0}^{n-1} u_j\right) \tag{4.5.13}$$

has an exponentially stable fixed point at the origin. Then (4.5.11) *has a fixed point that is asymptotically stable.*

Proof. Let us say $y_n = \|x_{n+1} - x_n\|$. Then we have $\|x_{n+1} - x_0\| \leq \sum_{j=0}^{n} \|x_{j+1} - x_j\| \leq \sum_{j=0}^{n} y_j$. Since g is nondecreasing, it follows that

$$y_{n+1} \leq g\left(y_n, \sum_{j=0}^{n} y_j, \sum_{j=0}^{n-1} y_j\right).$$

By Theorem 1.6.6 we then obtain

$$y_n \leq u_n,$$

where u_n is the solution of (4.5.13), provided that $y_0 \leq u_0$. If the origin is exponentially stable for (4.5.13), it follows that for suitable u_0 the sequence u_n will tend to zero and the same will happen to $y_n = \|x_{n+1} - x_n\|$. Moreover for all $p \geq 0$

$$\|x_{n+p} - x_n\| \leq \sum_{j=1}^{p} y_{n+j}.$$

Now by exponential stability of the origin of (4.5.13) and by Theorem 4.1.1 it follows that it is also l_1-stable. Then the series $\sum y_j$ is convergent and then for suitable n, $\sum_{j=1}^{p} y_{n+j}$ can be taken arbitrarily small, showing that x_k is a Cauchy sequence. ■

4.6. Variation of Constants

Consider the equation

$$y_{n+1} = A(n)y_n + f(n, y_n), \qquad y_{n_0} = y_0, \tag{4.6.1}$$

where $A(n)$ is an $s \times s$ nonsingular matrix and $f: N_{n_0}^{+} \times R^s \to R^s$.

Theorem 4.6.1. *The solution $y(n, n_0, y_0)$ of (4.6.1) satisfies the equation*

$$y_n = \Phi(n, n_0)y_{n_0} + \sum_{j=n_0}^{n-1} \Phi(n, j+1)f(j, y_j), \tag{4.6.2}$$

where $\Phi(n, n_0)$ is the fundamental matrix of the equation

$$x_{n+1} = A(n)x_n. \tag{4.6.3}$$

Proof. Let

$$y(n, n_0, y_0) = \Phi(n, n_0)x_n, \qquad x_{n_0} = y_0. \tag{4.6.4}$$

Then substituting in (4.6.1), we get

$$\Phi(n+1, n_0)x_{n+1} = A(n)\Phi(n, n_0)x_n + f(n, y_n)$$

from which we see that

$$\Delta x_n = \Phi^{-1}(n+1, n_0)f(n, y_n),$$

and

$$x_n = \sum_{j=n_0}^{n-1} \Phi(n_0, j+1)f(j, y_j) + x_{n_0}. \qquad ■$$

From (4.6.4), it follows that

$$y(n, n_0, y_0) = \Phi(n, n_0)y_0 + \sum_{j=n_0}^{n-1} \Phi(n, j+1)f(j, y_j). \tag{4.6.5}$$

Consider now the equation

$$x_{n+1} = f(n, x_n). \tag{4.6.6}$$

Lemma 4.6.2. *Assume that* $f: \mathbf{N}_{n_0}^+ \times \boldsymbol{R}^s \to \boldsymbol{R}^s$ *and* f *possesses partial derivatives on* $\mathbf{N}_{n_0}^+ \times \boldsymbol{R}^s$. *Let the solution* $x(n) \equiv x(n, n_0, x_0)$ *of* (4.6.6) *exist for* $n \geq n_0$ *and let*

$$H(n, n_0, x_0) = \frac{\partial f(n, x(n, n_0, x_0))}{\partial x}. \tag{4.6.7}$$

Then

$$\Phi(n, n_0, x_0) = \frac{\partial x(n, n_0, x_0)}{\partial x_0} \tag{4.6.8}$$

exists and is the solution of

$$\Phi(n+1, n_0, x_0) = H(n, n_0, x_0)\Phi(n, n_0, x_0), \tag{4.6.9}$$

$$\Phi(n_0, n_0, x_0) = I. \tag{4.6.10}$$

Proof. By differentiating (4.6.6) with respect to x_0 we have

$$\frac{\partial x_{n+1}}{\partial x_0} = \frac{\partial f}{\partial x_n}\frac{\partial x_n}{\partial x_0}.$$

Then (4.6.9) follows from the definition of Φ.

We are now able to generalize Theorem 4.6.1 to the equation

$$y_{n+1} = f(n, y_n) + F(n, y_n). \quad \blacksquare \tag{4.6.11}$$

Theorem 4.6.2. *Let* $f, F: \mathbf{N}_{n_0}^+ \times \boldsymbol{R}^s \to \boldsymbol{R}^s$ *and let* $\partial f/\partial x$ *exist and be continuous and invertible on* $\mathbf{N}_{n_0}^+ \times \boldsymbol{R}^s$. *If* $x(n, n_0, x_0)$ *is the solution of*

$$x_{n+1} = f(n, x_n), \qquad x_{n_0} = x_0, \tag{4.6.12}$$

then any solution of (4.6.11) *satisfies the equation*

$$y(n, n_0, x_0) = x\left(n, n_0, x_0 + \sum_{j=n_0}^{n-1} \psi^{-1}(j+1, n_0, v_j, v_{j+1})F(j, y_j)\right) \tag{4.6.13}$$

where

$$\psi(n, n_0, v_j, v_{j+1}) = \int_0^1 \Phi(n, n_0, sv_j + (1-s)v_{j-1})\, ds$$

and v_j satisfies the implicit equation (4.6.14). ■

Proof. Let us put $y(n, n_0, x_0) = x(n, n_0, v_n)$ and $v_0 = x_0$. Then

$$\begin{aligned} y(n+1, n_0, x_0) &= x(n+1, n_0, v_{n+1}) - x(n+1, n_0, v_n) + x(n+1, n_0, v_n) \\ &= f(n, x(n, n_0, v_n)) + F(n, x(n, n_0, v_n)) \end{aligned}$$

from which we get

$$x(n+1, n_0, v_{n+1}) - x(n+1, n_0, v_n) = F(n, y_n).$$

Applying the mean value theorem we have

$$\int_0^1 \frac{\partial x(n+1, n_0, sv_{n+1} + (1-s)v_n)}{\partial x_0}\, ds(v_{n+1} - v_n) = F(n, y_n)$$

and hence by (4.6.8)

$$\int_0^1 \Phi(n+1, n_0, sv_{n+1} + (1-s)v_n)\, ds(v_{n+1} - v_n) = F(n, y_n),$$

which is equivalent to

$$\psi(n+1, n_0, v_n, v_{n+1})(v_{n+1} - v_n) = F(n, y_n). \tag{4.6.14}$$

It now follows that

$$v_{n+1} - v_n = \psi^{-1}(n+1, n_0, v_n, v_{n+1})F(n, y_n)$$

and

$$v_n = v_0 + \sum_{j=n_0}^{n-1} \psi^{-1}(j+1, n_0, v_j, v_{j+1})F(j, y_j) \tag{4.6.15}$$

from which the conclusion results. ■

Corollary 4.6.1. *Under the hypothesis of Theorem* 4.6.2, *the solution $y(n, n_0, x_0)$ can be written in the following form.*

$$\begin{aligned} y(n, n_0, x_0) = x(n, n_0, x_0) &+ \psi(n, n_0, v_n, x_0) \\ &\cdot \sum_{j=n_0}^{n-1} \psi^{-1}(j+1, n_0, v_j, v_{j+1})F(j, v_j) \end{aligned} \tag{4.6.16}$$

Proof. Apply the mean value theorem once more to (4.6.13). ■

Corollary 4.6.2. *If $f(n, x) = A(n)x$, then* (4.6.16) *reduces to* (4.6.2).

Proof. In this case $x_n = \Phi(n, n_0)x_0$ and

$$\Phi(n, n_0, x_0) \equiv \psi(n, n_0) \equiv \psi(n, n_0, v_n, v_{n+1}),$$

and therefore, we have $v_{n+1} - v_n = \psi^{-1}(n+1, n_0, v_n, v_{n+1})F(n, y_n)$ and $v_n = x_0 + \sum_{j=n_0}^{n-1} \psi^{-1}(j+1, n_0, v_j, v_{j+1})F(j, y_j)$ from which follows the claim. ■

4.7. Stability by First Approximation

Consider the equation

$$y_{n+1} = A(n)y_n + f(n, y_n), \tag{4.7.1}$$

where $y \in \boldsymbol{R}^s$, $A(n)$ is an $s \times s$ matrix, $f: \mathbf{N}_{n_0} \times B_a \to B_a$ and $f(n, 0) = 0$.

When f is small in the sense to be specified, one can consider (4.7.1) as a perturbation of the equation

$$x_{n+1} = A(n)x_n \tag{4.7.2}$$

and the question arises whether the properties of stability of (4.7.2) are preserved for (4.7.1). The following theorems offer an answer to such a question.

Theorem 4.7.1. *Assume that*

$$\|f(n, y_n)\| \leq g_n \|y_n\|, \tag{4.7.3}$$

where g_n are positive and $\sum_{n=n_0}^{\infty} g_n < \infty$. Then if the zero solution of (4.7.2) is uniformly stable (or uniformly asymptotically stable), then the zero solution of (4.7.1) is uniformly stable (or uniformly asymptotically stable).

Proof. By (4.6.2) we get

$$y_n = \Phi(n, n_0)y_0 + \sum_{j=0}^{n-1} \Phi(n, j+1)f(j, y_j).$$

Because of Theorem 4.2.1 we have, using (4.7.3),

$$\|y_n\| \leq M\|y_0\| + M \sum_{j=n_0}^{n-1} g_j \|y_j\|.$$

Corollary 1.6.2 yields

$$\|y_n\| \leq M\|y_0\| \exp\left(M \sum_{j=n_0}^{n-1} g_j\right),$$

from which follows the proof, provided that x_0 is small enough such that $M\|y_0\| \exp(M \sum_{j=n_0}^{\infty} g_j) < a$.

In the case of uniform asymptotic stability, it follows that for $n > N$, $\|\Phi(n, n_0)y_0\| < \varepsilon$, for every $\varepsilon > 0$, the previous inequality can be written

$$\|y_n\| \leq \varepsilon \exp\left(M \sum_{n_0}^{\infty} g_j\right)$$

from which we conclude that $\lim y_n = 0$. ■

Corollary 4.7.1. *If the matrix A is constant such that the solutions of* (4.7.2) *are bounded, then the solutions of the equation*

$$y_{n+1} = (A + B(n))y_n \tag{4.7.4}$$

are bounded provided that

$$\sum_{n=n_0}^{\infty} \|B(n)\| < \infty. \tag{4.7.5}$$

Theorem 4.7.2. *Assume that*

$$\|f(n, y_n)\| \leq L\|y_n\|, \tag{4.7.6}$$

where $L > 0$ sufficiently small, and the solution $x_n = 0$ of (4.7.2) *is uniformly asymptotically stable. Then the solution $y_n = 0$ of* (4.7.1) *is exponentially asymptotically stable.*

Proof. By using the result of Theorem 4.2.2, we have

$$\|\phi(n, n_0)\| < H\eta^{n-n_0}, \qquad H > 0, \qquad 0 < \eta < 1,$$

and because of (4.7.6), we get

$$\|y_n\| \leq H\eta^{n-n_0}\|y_0\| + LH\eta^{n-1} \sum_{j=n_0}^{n-1} \eta^{-j}\|y_j\|.$$

By introducing the new variable $p_n = \eta^{-n}\|y_n\|$, we see that

$$p_n \leq H\eta^{-n_0}\|y_0\| + LH\eta^{-1} \sum_{j=n_0}^{n-1} p_j.$$

Using Corollary 1.6.2 again, we arrive at

$$\begin{aligned} p_n &\leq H\eta^{-n_0}\|y_0\| \prod_{j=n_0}^{n-1} (1 + LH\eta^{-1}) \\ &= H\eta^{-n_0}\|y_0\|(1 + LH\eta^{-1})^{n-n_0}, \end{aligned}$$

which implies

$$\|y_n\| \leq H\|y_0\|(\eta + LH)^{n-n_0}. \tag{4.7.8}$$

If $\eta + LH < 1$, that is $L < \frac{1-\eta}{H}$ the conclusion follows. ■

Corollary 4.7.2. (*Perron*). *Consider the equation*

$$y_{n+1} = Ay_n + f(n, y_n), \tag{4.7.9}$$

where A has all the eigenvalues inside the unit disk and moreover

$$\lim_{y\to 0} \frac{\|f(n, y)\|}{\|y\|} = 0 \tag{4.7.10}$$

uniformly with respect to n, then the zero solution of (4.7.3) *is exponentially asymptotically stable.*

We can similarly prove the following result.

Theorem 4.7.3. *Assume that the zero solution of*

$$x_{n+1} = Ax_n,$$

where A is an $s \times s$ matrix, is asymptotically stable. If

$$\sum_{n=n_0}^{\infty} \|B(n)\| < +\infty$$

then the zero solution of (4.7.4) *is asymptotically stable.*

Next theorem concerns instability of the null solution. We shall merely state such a result for completeness leaving its proof as an exercise.

Theorem 4.7.4. *Assume that the zero solution of $x_{n+1} = Ax_n$ is unstable and $\lim_{n\to\infty} \frac{\|f(n, y_n)\|}{\|y_n\|} = 0$. Then the origin is unstable for* (4.7.1).

4.8. Lyapunov Functions

The most powerful method for studying the stability properties of a critical point is the Lyapunov's second method. It consists in the use of an auxiliary function, which generalizes the role of the energy in mechanical systems. For differential systems the method has been used since 1892, while for difference equations its use is much more recent. In order to characterize such auxiliary functions, we need to introduce a special class of functions.

Definition 4.8.1. A function ϕ is said to be of class K if it is continuous in $[0, a)$, strictly increasing and $\phi(0) = 0$.

It is easy to check that the product of any two functions of class K is in the same class and the inverse of such a function is in the same class. Let $V(n, x)$ be a function defined on $I_{n_0} \times B_a$, which assumes values in $\boldsymbol{R}^+$.

Definition 4.8.2. The function $V(n, x)$ is positive definite (or negative definite) if there exists a function $\phi \in K$ such that

$$\phi(\|x\|) \le V(n, x) \qquad (\text{or } V(n, x) \ge -\phi(\|x\|))$$

for all $(n, x) \in \boldsymbol{N}^+_{n_0} \times B_a$.

Definition 4.8.3. A function $V(n, x) \ge 0$ is said to be decrescent if there exists $\phi \in K$ such that

$$V(n, x) \le \phi(x)$$

for all $(n, x) \in \boldsymbol{N}^+_{n_0} \times B_a$.

Let us consider the equation

$$y_{n+1} = f(n, y_n), \tag{4.8.1}$$

where $f: \boldsymbol{N}^+_{n_0} \times B_a \to \boldsymbol{R}^s$, $f(n, 0) = 0$ and $f(n, x)$ is continuous in x. Let $y(n, n_0, y_0)$ be the solution of (4.8.1), having (n_0, y_0) as initial condition, which is defined for $n \in \boldsymbol{N}^+_{n_0}$.

We shall now consider the variation of the function V along the solutions of (4.8.1)

$$\Delta V(n, y_n) = V(n + 1, y_{n+1}) - V(n, y_n). \tag{4.8.2}$$

If there is a function $\omega: \boldsymbol{N}^+_{n_0} \times \boldsymbol{R} \to \boldsymbol{R}$ such that

$$\Delta V(n, y_n) \le \omega(n, V(n, y_n)),$$

then we shall consider the inequality

$$\begin{aligned} V(n + 1, y_{n+1}) &\le V(n, y_n) + \omega(n, V(n, y_n) \\ &\equiv g(n, V(n, y_n)) \end{aligned} \tag{4.8.3}$$

to which we shall associate the comparison equation

$$u_{n+1} = g(n, u_n) \equiv u_n + \omega(n, u_n). \tag{4.8.4}$$

The auxiliary functions $V(n, x)$ are called Lyapunov functions. In the following, we shall always assume that such functions are continuous with respect to the second argument.

Theorem 4.8.1. *Suppose there exists two function $V(n, x)$ and $g(n, u)$ satisfying the following conditions:*

(1) $g : \mathbf{N}_{n_0}^+ \times \boldsymbol{R}^+ \to \boldsymbol{R}$, $g(n, 0) = 0$, *$g(n, u)$ is nondecreasing in u;*
(2) $V : \mathbf{N}_{n_0}^+ \times B_a \to \boldsymbol{R}^+$, $V(n, 0) = 0$, *and $V(n, x)$ is positive definite and continuous with respect to the second argument;*
(3) *V satisfies* (4.8.3).

Then

(a) *the stability of $u = 0$ for* (4.8.4) *implies the stability of $y_n = 0$;*
(b) *the asymptotic stability of $u = 0$ implies the asymptotic stability of $y_n = 0$.*

Proof. By Theorem 1.6.1, we know that

$$V(n, y_n) \le u_n, \qquad n \in \mathbf{N}_{n_0}^+,$$

provided that $V(n_0, y_0) \le u_0$. From the hypothesis of positive definiteness we obtain for $\phi \in K$,

$$\phi(\|y_n\|) \le V(n, y_n) \le u_n.$$

If the zero solution of the comparison equation is stable, we get that $u_n < \phi(\varepsilon)$ provided that $u_0 < \eta(\varepsilon, n_0)$, which implies

$$\phi(\|y_n\|) \le V(n, y_n) < \phi(\varepsilon),$$

from which we get

$$\|y_n\| < \varepsilon. \tag{4.8.5}$$

By, using the hypothesis of continuity of V with respect to the second argument, it will be possible to find a $\delta(\varepsilon, n_0)$ such that $\|y_{n_0}\| < \delta(\varepsilon, n_0)$ will imply $V(n_0, y_{n_0}) \le u_0$.

In the case of asymptotic stability, from

$$\phi(\|y_n\|) \le V(n, y_n) \le u_n$$

we get $\lim_{n\to\infty} \phi(\|y_n\|) = 0$ and consequently $\lim_{n\to\infty} y_n = 0$. ■

Corollary 4.8.1. *If there exists a positive definite function $V(n, x)$ such that on $\mathbf{N}_{n_0}^+ \times B_a$, V is continuous with respect to x and moreover, $\Delta V_n \le 0$, then the zero solution of* (4.8.1) *is stable.*

Proof. In this case $\omega(n, u) \equiv 0$ and the comparison equation reduces to $u_{n+1} = u_n$, which has stable zero solution. ■

Theorem 4.8.2. *Assume that there exists two functions $V(n, x)$ and $\omega(n, u)$ satisfying conditions (1), (2), (3) of the previous theorem and moreover suppose that V is decrescent. Then*

(*a*) *uniform stability of $u = 0$ implies uniform stability of $y_n = 0$.*
(*b*) *uniform asymptotic stability of $u = 0$ implies uniform asymptotic stability of $y_n = 0$.*

Proof. The proof proceeds as in the previous case except that we need to show that $\delta(\varepsilon, n_0)$ can be chosen independent on n_0. This can be done by using the hypothesis that $V(n, x)$ is decrescent, because in this case there exists a $\mu \in K$ such that $V(n, y_n) \le \mu(\|y_n\|)$. In fact, as before, we have

$$\phi(\|y_n\|) \le V(n, y_n) \le u_n$$

provided that $V(n_0, y_{n_0}) \le u_0 \le \eta(\varepsilon)$. If we take $\mu(\|y_{n_0}\|) = u_0 \le \eta(\varepsilon)$, that is $\|y_{n_0}\| < \delta(\varepsilon) \equiv \mu^{-1}(\eta(\varepsilon))$, then $\|y_n\| < \varepsilon$ for all $n \ge n_0$.

The uniform asymptotic stability will follow similarly. ■

Corollary 4.8.2. *If there exists a positive definite and decrescent function V such that*

$$\Delta V(n, y_n) \le 0,$$

then $y_n = 0$ is uniformly stable.

Corollary 4.8.3. *If there exists a function V such that*

$$\phi(\|y_n\|) \le V(n, y_n) \le \mu(\|y_n\|)$$

and

$$\Delta V(n, y_n) \le -\nu(\|y_n\|),$$

where ϕ, μ, $\nu \in K$ and the function $\omega \equiv \nu\mu^{-1}$ is a Lipshitz function with constant less than one then $y_n = 0$ is uniformly asymptotically stable.

Proof. Clearly $\|y_n\| \ge \mu^{-1}(V(n, y_n))$ and by substitution we have

$$\Delta V(n, y_n) \le -\nu(\mu^{-1}(V(n, y_n))) \equiv -\omega(V(n, y_n)).$$

The comparison equation is now

$$u_{n+1} = u_n - \omega(u_n), \tag{4.8.6}$$

where the righthand side is positive and nondecreasing, because of the hypothesis. The same hypothesis shows that the origin is asymptotically stable for (4.8.6) (see Problem 4.11) and because the equation is autonomous, it follows that the stability is also uniform.

With minor changes, one can show that the condition on ΔV can be substituted by $\Delta V(n, y_n) \leq -\nu(\|y_{n+1}\|)$.

If in Theorem 4.8.2 the condition that V is decrescent is removed, the asymptotic stability will remain. ∎

Theorem 4.8.3. *Assume that there exists a function V such that*

(1) $V: N_{n_0}^+ \times B_a \to R^+$; $V(n, 0) = 0$; *V is positive definite and continuous with respect to the second argument*;
(2) $\Delta V(n, y_n) \leq -\mu(\|y_n\|)$, $\mu \in K$.

Then the origin is asymptotically stable.

Proof. By Theorem 4.8.1 we know that the origin is stable. Suppose it is not asymptotically stable, then there exists a solution $y(n, n_0, y_0)$ and a set $J_{n_0} \subset N_{n_0}^+$ such that $\|y(n, n_0, y_0)\| > \varepsilon > 0$. We then have

$$V(n+1, y(n+1, n_0, y_0)) \leq V(n, y_n) - \mu(\varepsilon),$$

if $n \in J_{n_0}$ or, since V is decreasing, $V(n+1, y_{n+1}) \leq V(n, y_n)$. Summing we have

$$V(n+1, y(n+1, n_0, y_0)) \leq V(n_0, y_{n_0}) - k\varepsilon,$$

where k is the number of elements in J_{n_0}. Taking the limit we get

$$\lim_{n\to\infty} V(n, y_n) = -\infty,$$

which contradicts the hypothesis that V is positive definite. ∎

The next theorem concerns the l_p stability and is related to the previous one.

Theorem 4.8.4. *Assume that there exists a function V such that*

(1) $V: N_{n_0}^+ \times B_a \to R^+$, $V(n, 0) = 0$, *positive definite and continuous with respect to the second argument*;
(2) $\Delta V(n, y_n) \leq -c\|y_n\|^p$,

where p, c are positive constant. Then $y_n = 0$ is l_p stable.

Proof. By Theorem 4.8.1 we know that $y_n = 0$ is stable, that means that for $n \geq n_0$, $\varepsilon \geq 0$, there exists $\delta(\varepsilon, n_0)$ such that for $\|y_{n_0}\| < \delta$, $\|y_n\| < \varepsilon$. Let us define the function $G(n) = V(n, y_n) + c\sum_{j=n_0}^{n-1} \|y_j\|^p$. Then

$$\Delta G(n) = \Delta V(n, y_n) + c\|y_n\|^p \leq 0.$$

Therefore $G(n) \leq G(n_0) = V(n_0, y_{n_0})$ for $n \geq n_0$ and

$$0 \leq G(n) = V(n, y_n) + c \sum_{j=n_0}^{n-1} \|y_n\|^p \leq V(n_0, y_{n_0})$$

from which it follows

$$\sum_{j=n_0}^{n-1} \|y_n\|^p \leq \frac{1}{c} V(n_0, y_{n_0})$$

and

$$\sum_{j=n_0}^{\infty} \|y_n\|^p < \infty. \qquad \blacksquare$$

The next theorem is a generalization of the previous one. It is the discrete analog of La Salle's invariance principle. Let us consider the solution $y(n, n_0, y_{n_0})$ of

$$y_{n+1} = f(n, y_n), \qquad y_{n_0} = y_0, \tag{4.8.7}$$

and suppose that it is continuous with respect to the initial vector y_0.

Theorem 4.8.5. *Suppose that, for $y \in D \subset \boldsymbol{R}^s$,*

(1) *there exist two real valued functions $V(n, y)$, $\omega(y) \geq 0$ both continuous in y, with $V(n, y)$ bounded below such that*

$$\Delta V(n, y_n) \leq -\omega(y_n) \qquad \textit{for } n \geq n_0; \tag{4.8.8}$$

(2) *$y(n, n_0, y_{n_0}) \in D$ for $n \geq n_0$.*

Then either $y(n, n_0, y_{n_0})$ is unbounded or it approaches the set

$$E = \{x \in \bar{\boldsymbol{D}} \,|\, \omega(x) = 0\}. \tag{4.8.9}$$

Proof. By assumption $y_n \in \boldsymbol{D}$ for $n \geq n_0$ and $V(n, y_n)$ is decreasing along it. Because V is bounded below, it must approach a limit for $n \to \infty$ and $\omega(y_n)$ must approach zero. Then either the limit is finite and must lie in E, or it is infinite. $\blacksquare$

Corollary 4.8.4. *Suppose that $u(x)$ and $v(x)$ are continuous real valued functions such that*

$$u(x) \leq V(n, x) \leq v(x) \tag{4.8.10}$$

for $n \geq n_0$. Fixing η, consider the set $\boldsymbol{D}(\eta) = \{x \,|\, u(x) < \eta\}$, $\boldsymbol{D}_1(\eta) = \{x \,|\, v(x) < \eta\}$. Under the hypothesis of Theorem 4.8.5 with $\boldsymbol{D} \equiv \boldsymbol{D}(\eta)$, all solutions that start in $\boldsymbol{D}_1(\eta)$ remain in $\boldsymbol{D}(\eta)$ and approach E for $n \to \infty$.

Proof. Let $y_0 \in \boldsymbol{D}_1(\eta)$, then

$$u(y_n) \le V(n, y_n) \le V(n_0, y_0) \le v(y_0) < \eta$$

showing that $u(y_n) < \eta$ for $n \ge n_0$. ■

Example. Consider the equation

$$y_{n+1} = \boldsymbol{M}(n, y_n) y_n, \tag{4.8.11}$$

where $\boldsymbol{M}$ is an $s \times s$ matrix. Define $V(n, y_n)$ as

$$V(n, y_n) = \|y_n\| \tag{4.8.12}$$

then

$$\Delta V = \|\boldsymbol{M}(n, y_n) y_n\| - \|y_n\| \le (\|\boldsymbol{M}(n, y_n)\| - 1\|y_n\|.$$

Let $u(y_n) \equiv v(y_n) \equiv V(n, y_n)$. Then $\boldsymbol{D}(\eta) \equiv \boldsymbol{D}_1(\eta) = \{y \mid \|y\| < \eta\}$. For all $y \in \boldsymbol{D}$, let $\|\boldsymbol{M}(n, y)\| < \alpha(y)$ and $\omega(y) = (1 - \alpha(y)\|y\|$. It then follows that

$$\Delta V \le -\omega(y).$$

If $\alpha(y) < 1$, for $y \in \boldsymbol{D}(\eta)$, $\omega(y)$ is positive. The set E is the origin and possibly something on the boundary of $\boldsymbol{D}(\eta)$. Because V is decreasing on $\boldsymbol{D}(\eta)$, it follows that this last possibility cannot occur. Then the solution starting in $\boldsymbol{D}(\eta)$ cannot leave this set and it will tend to the origin. $\boldsymbol{D}(\eta)$ is a domain of asymptotic stability of the origin. Different choices of η and of the norm used will give different domains of asymptotic stability. Of course, the union of all these domains is still a domain of asymptotic stability.

If $\boldsymbol{M}$ is independent on n with spectral radius less than one, then it is possible to choose a vector norm such that $\alpha(0) < 1$ with $\alpha(x)$ continuous. The previous result shows that it is possible to define a nonempty domain of asymptotic stability.

Definition 4.8.4. The positive limit set $\Omega(y_{n_0})$ of a sequence y_n, $n \in \boldsymbol{N}_{n_0}^+$ is the set of all the limit points of the sequence. That is, $y \in \Omega(y_{n_0})$ if there exists an unbounded subset $J_{n_0} \subset \boldsymbol{N}_{n_0}^+$ such that $y_{n_i} \to y$ for $n_i \in J_{n_0}$.

Definition 4.8.5. A set $\boldsymbol{S} \subset \boldsymbol{R}^s$ is said invariant if for $y_0 \in \boldsymbol{S}$ follows that $y(n, n_0, y_0) \in \boldsymbol{S}$.

In the case of autonomous difference equation

$$y_{n+1} = f(y_n), \tag{4.8.13}$$

where f is a continuous vector valued function with $f(0) = 0$. Theorem 4.8.5 assumes the following form.

Theorem 4.8.6. *Suppose that for $y \in \boldsymbol{D} \subset \boldsymbol{R}^s$,*

(1) *there exist two real valued functions $V(y)$, $\omega(y)$ both continuous in y, with V bounded below, and*

$$\Delta V(y_n) \leq -\omega(y_n);$$

(2) *$y_n \in \boldsymbol{D}$ for $n \geq n_0$;*

then either y_n is unbounded or approaches the maximum invariant set M contained in E.

Proof. Same as before and moreover, the positive limit set of any bounded solution of an autonomous equation is nonempty, invariant and compact (see problem 4.12). ■

Corollary 4.8.5 can be rewritten as follows.

Corollary 4.8.6. *If in Theorem 4.8.6 the set $\boldsymbol{D}$ is of the form*

$$\boldsymbol{D}(\eta) = \{x \mid V(x) < \eta\}$$

for some $\eta > 0$, then all the solutions that start in $\boldsymbol{D}(\eta)$ remain in it and approach M as $n \to \infty$.

The next result, again for autonomous equations, imposes conditions on the second differences of the Lyapunov function.

Theorem 4.8.7. *Suppose that $V: \boldsymbol{R}^s \to \boldsymbol{R}$ is a continuous function with $\Delta^2 V(y_n) > 0$ for $y_n \neq 0$. Then for any $y_0 \in \boldsymbol{R}^s$, either $y(n, y_0)$ is unbounded or it tends to zero for $n \to \infty$. Likewise if $\Delta^2 V(y_n) < 0$, $y_n \neq 0$.*

Proof. Put $\Delta^2 V > 0$ in the form

$$V(y_{n+2}) - V(y_{n+1}) > V(y_{n+1}) - V(y_n).$$

If there exists a $k \in \boldsymbol{N}^+_{n_0}$ such that $V(y_{k+1}) - V(y_k) \geq 0$, then $V(y_{n+1}) - V(y_n) > 0$ for $n > k$, otherwise we get $V(y_{n+1}) < V(y_n)$ for all n. In both cases V is a monotone function. Suppose that V is not increasing. Consider the positive limit set $\Omega(y_0)$ of $y(n, n_0, y_0)$. If $\Omega(y_0)$ is empty then $y(n, n_0, y_0)$ is unbounded and the theorem is proved. If $\Omega(y_0)$ is not empty, $V(y_n)$ must be constant on $\Omega(y_0)$ because the limit of a monotone function is unique. But this is impossible because $\Delta^2 V > 0$, unless $\Omega(y_0) = \{0\}$. The other cases are proved similarly. ■

4.9. *Domain of Asymptotic Stability*

All the previous results say that if the initial value is small enough, then the origin has some kind of stability. In applications one is more interested in the domain of asymptotic stability since one needs to know where to start with the iterations. In other words, one needs to know the domain in $\boldsymbol{R}^s$, containing the initial values, starting from which solutions will eventually tend to the fixed point. This problem is a difficult one and will also be discussed in the next chapter. We shall give some results in this direction in the case of autonomous difference equations

$$y_{n+1} = f(y_n), \tag{4.9.1}$$

where $f \in \boldsymbol{C}[B_a, \boldsymbol{R}^s]$, and $f(0) = 0$.

Theorem 4.9.1. *Suppose that there exists a continuous function $V: B_a \to \boldsymbol{R}^+$, $V(0) = 0$ and $\Delta V(y_n) < 0$. Then the origin is asymptotically stable. Moreover if $B_a \equiv \boldsymbol{R}^s$ and $V(y) \to \infty$ as $y \to \infty$, then the origin is globally asymptotically stable.*

Proof. The proof of stability follows from the same arguments used in Theorem and Corollary 4.8.1, except that continuity of V is used instead of positive definiteness of V.

For proving asymptotic stability, we observe that for $y_n \in \boldsymbol{B}_\varepsilon$, where $0 < \varepsilon < \alpha$, $V(y_n)$ is strictly decreasing and it must converge to zero. Again by continuity of V, the sequence y_n itself must converge to zero.

Now suppose that the last hypothesis is true. Then it is clear that we are not restricted to take $\|y_n\| < \alpha$ and hence the proof is complete. ∎

As an example, consider (4.9.1) when f is linear,

$$y_{n+1} = Ay_n \tag{4.9.2}$$

and let us take

$$V(y_n) = y_n^T B y_n, \tag{4.9.3}$$

where B is a symmetric positive definite matrix. The demand that $\Delta V(y_n) < 0$ becomes

$$y_n^T A^T B A y_n - y_n^T B y_n < 0$$

that is, we must have

$$A^T BA - B = -C, \tag{4.9.4}$$

where C is any positive definite matrix. This leads to the following result.

Corollary 4.9.1. *If there is a symmetric positive definite matrix B such that (4.9.4) is verified, then the origin is asymptotically stable.*

The converse of this corollary is also true.

Theorem 4.9.2. *Suppose* (4.9.2) *is asymptotically stable, then there exists two positive definite matrices B and C such that* (4.9.4) *is verified.*

The statement analogous to (4.9.4) in the continuous case is

$$S^T G + GS = -G_1, \tag{4.9.5}$$

where G and G_1 are symmetric positive definite matrices, and S has the eigenvalues with negative real part. Equation (4.9.5) is called Lyapunov matrix equation. There is, of course, a correspondence between (4.9.4) and (4.9.5). In fact by putting $S = (A+I)(A-I)^{-1}$, (4.9.5) is transformed into an equation of the form (4.9.4) and vice versa by putting $A = (I-S)^{-1}(I+S)$. To find the matrix B, one needs to solve the matrix equation (4.9.4) where A is given and C is chosen appropriately. The methods of solution of the equation (4.9.4) (or alternatively 4.9.5) have been studied extensively in the last ten years.

The following theorem also gives the region of the asymptotic stability for the zero solution of (4.9.1) and it is the discrete version of the Zubov's theorem.

Theorem 4.9.3. *Assume that there exists two functions V and ϕ satisfying the following conditions*:

(1) $V: \boldsymbol{C}[\boldsymbol{R}^s, \boldsymbol{R}^+]$, $\quad V(0) = 0$, $\quad V(x) > 0 \quad$ *for* $x \neq 0$,
(2) $\phi: \boldsymbol{C}[\boldsymbol{R}^s, \boldsymbol{R}^+]$, $\quad \phi(0) = 0$, $\quad \phi(x) > 0 \quad$ *for* $x \neq 0$,
(3) $V(y_{n+1}) = (1+\phi(y_n))V(y_n) - \phi(y_n)$.

Then $D = \{x \mid V(x) < 1\}$ *is the domain of asymptotic stability.*

Proof. By condition (3), which can be written as

$$1 - V(y_{n+1}) = (1+\phi(y_n))(1 - V(y_n)),$$

one obtains

$$1 - V(y_n) = \prod_{j=0}^{n-1} (1+\phi(y_j))(1 - V(y_0)). \tag{4.9.6}$$

Suppose now that y_0 lies in the region of asymptotic stability. Then, the left-hand side tends to one by hypothesis, the right-hand side will also be convergent and will tend to $C(1 - V(y_0))$, where $C = \prod_{j=0}^{\infty} (1+\phi(y_j)) > 1$. It follows then that $V(y_0) = (C-1)/C < 1$, which shows that $y_0 \in \boldsymbol{D}$. Conversely if $y_0 \notin \boldsymbol{D}$, then again from (4.9.6) we see that $V(y_n)$ will remain always outside $\boldsymbol{D}$ and V will never be zero. This implies that y_n will not tend to zero. ■

To use this theorem, one needs to know the solution y_j of (4.9.1), and then from (4.9.6) find the function V, which will define the set $\boldsymbol{D}$. Unfortunately this can be done only for some cases.

Theorems 4.8.5 and 4.8.7 and their corollaries also give the asymptotic stability domains in terms of $\boldsymbol{D}(\eta)$ for the set E. When the maximum invariant set contained in E, reduces to a point, Theorem 4.8.7 gives the asymptotic stability domain for a critical point.

Theorem 4.8.7 can also be used to obtain the domain of asymptotic stability. Suppose that the condition $\Delta^2 V > 0$ holds true only in an open region H containing the origin. Then put

$$\Delta \max = \max\{\Delta V(x) \mid x \in \text{boundary } H\}$$

and

$$E_j = \{x \in H \mid \Delta V(y(j, x)) > \Delta \max\}$$

for $j = 0, 1, 2, \ldots$, where $y(j, x) \equiv y(j, n_0, x)$.

Theorem 4.9.4. *If the regions E_j are bounded and nonempty, then they are domains of asymptotic stability for* (4.9.1).

Proof. If E_j is not empty and $x \in E_j$, $y(j+k, x) \in H$, since ΔV is not decreasing along any trajectory in E_j. Moreover

$$\Delta V(y(j+k+1, x)) - \Delta V(y(j+k, x)) = \Delta^2 V(y(j+k, x)) > 0$$

and $\Delta V(y(j+k, x)) > \Delta V(y(j, x)) \geq \Delta \max$. This means that $y(j+k, x) \in E_j$ for all k, from which we see that $y(n, x)$ is bounded. By Theorem 4.8.7 it follows that $y(n, x) \to 0$. ■

Similar results hold in the case where $\Delta^2 V < 0$ in H and the regions

$$F_j = \{x \in H \mid \Delta V(y(j, x))\} < \Delta \min.$$

$$\Delta \min = \min\{\Delta V(x) \mid x \in \text{boundary } H\}.$$

As an example, we consider the following system arising in biomathematics

$$x_{n+1} = y_n,$$

$$y_{n+1} = a x_n - y_n^2, |a| < 3^{-\frac{1}{2}}.$$

Consider $V(x, y) = \dfrac{2a^2x^2}{1+a^2} + y^2$. One obtains

$$\Delta^2 V(x, y) = \frac{y^2}{1+a^2}(1 - 3a^2 + 4ax - 2y^2) + (ay - (ax - y^2)^2)^2.$$

Here $H = \{(x, y) | 1 - 3a^2 + 4ax - 2y^2 > 0\}$ contains the origin. For $a = 1/4$, one shows that Δ max $= -0.0364$, and then

$$E_0 = \left\{(x, y) \middle| \left(y^2 - \frac{x}{4}\right)^2 - \frac{2x^2}{17} - \frac{15y^2}{17} > -0.0364\right\}.$$

4.10. Converse Theorems

This section will be devoted to the construction of Lyapunov functions when certain stability properties hold. In this construction of Lyapunov functions, however, one uses the solutions of the problem and this implies that this construction is of little use in practice. The real importance of converse theorems lies therefore in the fact that by means of such theorems, it is possible to prove some results on total stability, which is a very important concept in applications. For this reason we shall present only those results that we shall use later.

Theorem 4.10.1. *Suppose that the zero solution of* (4.8.1) *is uniformly stable, then there exists a function V, positive definite and decrescent, such that* $\Delta V \le 0$ *along the solutions.*

Proof. Consider the function

$$V(n, y_n) = \sup_{k \ge n} \|y(k, n, y_n)\|. \tag{4.10.1}$$

As usual, $y_n \equiv y(n, n_0, y_0)$. From (4.10.1) it is immediate that $V(n, y_n) \ge \|y_n\|$ showing that V is positive definite. From the definition of uniform stability we know that $\|y(k, n, y_n)\| < \varepsilon$ if $\|y_n\| < \delta(\varepsilon)$. Without loss of generality, we shall suppose that $\delta \in K$. (See problems 4.8 and 4.9.) Let $\phi = \delta^{-1}$. Then we can write $\|y(k, n, y_n)\| \le \phi(\|y_n\|)$, from which we obtain $V(n, y_n) \le \phi(\|y_n\|)$, showing that V is decrescent. On the other hand, for every solution one has

$$\sup_{k \ge n} \|y(k, n, y(n, n_0, y_0))\| = \sup_{k \ge n} \|y(k, n_0, y_0)\|$$

and

$$V(n+1, y_{n+1}) = \sup_{k \geq n+1} \|y(k, n_0, y_0)\| \leq \sup_{k \geq n} \|y(k, n_0, y_0)\|$$
$$= V(n, y_n)$$

and this completes the proof. ■

Theorem 4.10.2. *Suppose that the zero solution of* (4.8.1), *where* $f(n, x)$ *is locally Lipschitzian around the origin, is uniformly asymptotically stable, then there exists a function* V, *positive definite and decrescent, such that* $\Delta V(n, y_n) < -\mu(\|y_{n+1}\|)$ *along the solutions, and* V *is locally Lipschitzian.*

Proof. Consider a function $G(r)$ defined for $r > 0$, $G(0) = 0$, $G'(0) = 0$, $G'(r) > 0$, $G''(r) > 0$ and let $\alpha > 1$. Since $G(r) = \int_0^r du \int_0^u G''(\nu)\, d\nu$ and $G\left(\frac{r}{\alpha}\right) = \int_0^{r/\alpha} du \int_0^u G''(\nu)\, d\nu$, we have, setting $u = \omega/\alpha$, $G\left(\frac{r}{\alpha}\right) = \frac{1}{\alpha}\int_0^r d\omega \int_0^{\omega/\alpha} G''(\nu)\, d\nu < \frac{1}{\alpha}\int_0^r d\omega \int_0^{\omega} G''(\nu)\, d\nu = \frac{1}{\alpha} G(r)$. Define

$$V(n, y_n) = \sup_{k \geq 0} G(\|y(n+k, n, y_n)\|)\frac{1+\alpha k}{1+k}.$$

For $k = 0$, we have

$$G(\|y_n\|) \leq V(n, y_n).$$

The uniform stability of the origin implies that (see previous theorem) there exists a $\phi \in k$ such that

$$\|y(n+k, n, y_n)\| < \phi(\|y_n\|)$$

and therefore

$$G(\|y(n+k, n, y_n)\|) < G(\phi(\|y_n\|)).$$

Consequently, since $\frac{1+\alpha k}{1+k} < \alpha$, $V(n, y_n) \leq \alpha G(\phi\|y_n\|)$. Furthermore, asymptotic stability implies that for $k \geq \mathbf{N}(\varepsilon)$, $\|y(n+k, n, y_n)\| < \varepsilon$, and hence, we get $\|y(n+k, n, y_n)\| < \frac{\|y_n\|}{\alpha}$ for $k \geq \mathbf{N}\left(\frac{\|y_n\|}{\alpha}\right)$. Thus, $G(\|y(n+k, n, y_n)\|) < G\left(\frac{\|y_n\|}{\alpha}\right)$, which in turn leads to $G(\|y(n+k, n, y_n)\|)\frac{1+\alpha k}{1+k} \leq \alpha G\left(\frac{\|y_n\|}{\alpha}\right) < G(\|y_n\|) \leq V(n, y_n)$. This shows that it is sufficient to consider

$k \in \left(0, N\left(\frac{\|y_n\|}{\alpha}\right)\right)$ obtaining

$$V(n, y_n) = \sup_{0 \le k \le \mathbf{N}(\|y_n\|/\alpha)} G(\|y(n+k, n, y_n)\|) \frac{1+k\alpha}{1+k}. \tag{4.10.2}$$

Let k_1 be the index where the sup is achieved so that

$$V(n, y_n) = G(\|y(n+k_1, n, y_n)\|) \frac{1+\alpha k_1}{1+k_1}. \tag{4.10.3}$$

Then

$$\begin{aligned} V(n+1, y_{n+1}) &= G(\|y(n+1+k_1, n+1, y_{n+1})\|) \frac{1+\alpha k_1}{1+k_1} \\ &= G(\|y(n+1+k_1, n+1, y_{n+1})\|) \\ &\quad \times \frac{1+\alpha(k_1+1)}{1+(k_1+1)} \left[1 - \frac{\alpha - 1}{(1+k_1)(1+\alpha+\alpha k_1)}\right] \\ &\le V(n, y_n)\left[1 - \frac{\alpha-1}{(1+k_1)(1+\alpha(k_1+1))}\right], \end{aligned}$$

from which

$$\begin{aligned} V(n+1, y_{n+1}) - V(n, y_n) &\le -\frac{(\alpha-1)V(n, y_n)}{\left[1+\mathbf{N}\left(\frac{\|y_{n+1}\|}{\alpha}\right)\right]\left[1+\alpha+\alpha\mathbf{N}\left(\frac{\|y_{n+1}\|}{\alpha}\right)\right]} \\ &\le -\frac{(\alpha-1)G(\|y_n\|)}{\left[1+\mathbf{N}\left(\frac{\|y_{n+1}\|}{\alpha}\right)\right]\left[1+\alpha+\alpha N\left(\frac{\|y_{n+1}\|}{\alpha}\right)\right]} \\ &\le -\frac{(\alpha-1)G\left(\frac{1}{L_{\delta^0}}\|y_{n+1}\|\right)}{\left[1+\mathbf{N}\left(\frac{\|y_{n+1}\|}{\alpha}\right)\right]\left[1+\alpha+\alpha\mathbf{N}\left(\frac{\|y_{n+1}\|}{\alpha}\right)\right]} \\ &\le -\mu(\|y_{n+1}\|), \end{aligned}$$

where L_{δ^0} is the Lipschitz constant. The last inequality follows from $\|y_{n+1}\| = \|f(n, y_n)\| \le L_{\delta^0}\|y_n\|$.

The function $\mu \in K$, because it is strictly increasing being $\mathbf{N}\left(\frac{\|y_n\|}{\alpha}\right)$ a decreasing function, and $G(0) = 0$. To complete the proof, we must show that we can choose a function G such that V is Lipschitzian. By hypothesis of uniform asymptotic stability it follows that for $r > 0$, there exists a $\delta(r)$ such that for $y', y'' \in B_{\delta(r)}$ and $y(n, n_0, y'), y(n, n_0, y'') \in B_r$. Because f is Lipschitzian one has

$$\|y(n, n_0, y') - y(n, n_0, y'')\| \leq L_r^{n-n_0}\|y' - y''\|.$$

We let

$$q \leq \min(L_r, L_r^{-1}), \qquad q < 1, \tag{4.10.5}$$

and

$$G(r) = A\int_0^r q^{N^{\delta(s)/\alpha}}\, ds, \tag{4.10.6}$$

where $N(r)$ is the same function defined before. $G(r)$ satisfies the condition required because $N(r)$ is decreasing and $\lim_{r\to 0} N(r) = \infty$.

We have seen that

$$V(n, y') = G(\|y(n + k_1, n, y')\|)\frac{1 + \alpha k_1}{1 + k_1}, \tag{4.10.7}$$

where

$$0 \leq k_1 \leq N\left(\frac{\|y'\|}{\alpha}\right). \tag{4.10.8}$$

For simplicity, let us put $r_1 = \|y(n + k_1, n, y')\|$ and $r_2 = \|y(n + k_1, n, y'')\|$. Suppose $r_2 \leq r_1$. Then $0 \leq G(r_1) - G(r_2) \leq G'(r_1)(r_1 - r_2) \leq Aq^{N(\delta(r)/\alpha)}(r_1 - r_2)$. But $r_1 - r_2 \leq \|y(n + k_1, n, y') - y(n + k_1, n, y'')\| \leq L_y^{k_1}\|y' - y''\|$. By substituting, we have

$$0 \leq G(r_1) - G(r_2) \leq Aq^{N(\delta(r)/\alpha)} \cdot L_r^{k_1}\|y' - y''\| \leq A\|y' - y''\|. \tag{4.10.9}$$

The last inequality follows from (4.10.8) and (4.10.5). Multiplying (4.10.9) by $\frac{1 + \alpha k_1}{1 + k_1}$ we get

$$0 \leq V(n, y') - G(\|y(n + k_1, n, y'')\|)\frac{1 + \alpha k_1}{1 + k_1} \leq \alpha A\|y' - y''\|$$

from which, because of the fact $V(n, y'') > G(\|y(n + k_1, n, y'')\|)\frac{1 + \alpha k_1}{1 + k_1}$, we obtain $V(n, y') - V(n, y'') \leq \alpha A\|y' - y''\|$. By interchanging the roles of y' and y'' one gets similarly

$$V(n, y'') - V(n, y') \geq -\alpha A\|y' - y''\|,$$

which shows that

$$|V(n, y') - V(n, y'')| \leq \alpha A\|y' - y''\|,$$

proving the theorem. ■

The following theorem is the converse in the case of *lp*-stability.

Theorem 4.10.3. *Suppose that the trivial solution of* (4.10.1) *is lp-stable and* $\|y(n, n_0, y_0)\| \leq g_n\beta(\|y_0\|)$ *with* $\beta \in \boldsymbol{K}$ *and* $\sum_{n=n_0}^{\infty} g_n = g$. *Then there exists a function* $V: \boldsymbol{N}_{n_0}^{+} \times B_a \to \boldsymbol{R}^{+}$ *such that it is positive definite, decrescent and* $\Delta V(n, y_n) \leq -\|y_n\|^p$.

Proof. Let

$$V(n, y_n) = \sum_{k=0}^{\infty} \|y(n+k, n, y_n)\|^p.$$

It follows that

$$V(n, y_n) \geq \|y_n\|^p,$$

which shows that V is positive definite. Moreover, $V(n, y_n) \leq \beta^p(\|y_n\|)\sum_{n=n_0}^{\infty} g_n = g\beta^p(\|y_n\|)$ and

$$\begin{aligned}\Delta V(n, y_n) &= \sum_{k=0}^{\infty} \|y(n+1+k, n+1, y_{n+1})\|^p - \sum_{k=0}^{\infty} \|y(n+k, n, y_n)\|^p \\ &= \sum_{k=0}^{\infty} \|y(n+1+k, n_0, y_0)\|^p - \sum_{k=0}^{\infty} \|y(n+k, n_0, y_0)\|^p \\ &= -\|y(n, n_0, y_0)\|^p,\end{aligned}$$

which completes the proof. ■

4.11. *Total and Practical Stability*

Let us consider the equations

$$y_{n+1} = f(n, y_n) + R(n, y_n) \tag{4.11.1}$$

$$y_{n+1} = f(n, y_n), \tag{4.11.2}$$

where $\boldsymbol{R}$ is a bounded, Lipschitz function in B_a and $R(n, 0) = 0$. We shall consider (4.11.1) as a perturbation of equation (4.11.2). Suppose that the zero solution of (4.11.2) has some kind of stability property, we want to know under what conditions on R the zero solution preserves some stability.

Definition 4.11.1. The solution $y = 0$ of (4.11.2) is said to be totally stable (or stable with respect to permanent perturbations) if for every $\varepsilon > 0$, there exists two positive numbers $\delta_1 = \delta_1(\varepsilon)$ and $\delta_2 = \delta_2(\varepsilon)$ such that every solution $y(n, n_0, y_0)$ of (4.11.1) lies in B_ε for $n \geq n_0$, provided that

$$\|y_0\| < \delta_1$$

and

$$\|R(n, y_n)\| < \delta_2 \text{ for } y_n \in B_\varepsilon, \qquad n \geq n_0.$$

Theorem 4.11.1. *Suppose that the trivial solution of* (4.11.2) *is uniformly asymptotically stable and moreover*

$$\|f(n, y') - f(n, y'')\| \leq L_r \|y' - y''\|, \qquad y', y'' \in B_r \subset B_a.$$

Then it is totally stable.

Proof. Let $\tilde{y}(n, n_0, y_0)$ be the solution of the unperturbated equation. The hypothesis of uniform asymptotic stability implies that for $0 < \delta_0 < a$, there exist $\delta(\delta_0) > 0$, $\delta(\delta_0) < a$ such that $y_n \in B_{\delta(\delta_0)}$, $\lim y_n = 0$ and moreover there exist the functions $a, b, c \in K$ and V such that for $n \in N_{n_0}^+$,

$$\text{(a)} \quad a(\|\tilde{y}\|) \leq V(n, \tilde{y}) \leq b(\|\tilde{y}\|),$$

$$\text{(b)} \quad \Delta V(n, \tilde{y}_n) \leq -c(\|\tilde{y}_{n+1}\|),$$

and

$$\text{(c)} \quad |v(n, y') - V(n, y'')| \leq M|y' - y''|,$$

$$\text{for } y', \qquad y'' \in B_{\delta(\delta_0)}, \qquad M > 0'$$

Let $0 < \varepsilon < \delta(\delta_0)$. Choose $\delta_1 > 0$, $\delta_2 > 0$ such that

$$b(2\delta_1) < a(\varepsilon), \qquad \delta_2 \leq \delta_1, \qquad \delta_2 < \frac{c(\delta_1)}{M}. \tag{4.11.3}$$

Suppose that ε is sufficiently small so that $L_\varepsilon \varepsilon + \delta_2 < \delta(\delta_0)$. Let $\|\boldsymbol{R}(n, y)\| < \delta_2$ for $y \in B_\varepsilon$. One then finds for $\|y\| \leq \varepsilon$, $\|\tilde{y}(n+1, n, y)\| = |f(n, y)| \leq L_\varepsilon \varepsilon$, $\|y(n+1, n, y) - \tilde{y}(n+1, n, y)\| = \|R(n, y)\| < \delta_2$, and $\|y(n+1, n, y)\| \leq L_\varepsilon \varepsilon + \delta_2 < \delta(\delta_0)$. Thus, for $\|y\| < \varepsilon$, we have

$$\begin{aligned} V(n+1, y(n+1, n, y)) - V(n, y) &= V(n+1, \tilde{y}(n+1, n, y)) - V(n, y) \\ &\quad + V(n+1, y(n+1, n, y)) \\ &\quad - V(n+1, \tilde{y}(n+1, n, y)) \\ &\leq -c(\|\tilde{y}(n+1, n, y))\| \\ &\quad + M\|y(n+1, n, y) \\ &\quad - \tilde{y}(n+1, n, y)\| \\ &\leq -c(\|\tilde{y}(n+1, n, y)\|) + M\delta_2 \\ &< -c(\|\tilde{y}(n+1, n, y)\|) \\ &\quad + c(\delta_1). \end{aligned} \tag{4.11.4}$$

Now suppose that there is an index $n_1 \geq n_0$ and a $y'_1 \in B_{\delta^1}$ such that $y(n_1, n_0, y') \notin B_\varepsilon$ and $y(n, n_0, y') \in B_\varepsilon$ for $n < n_1$. It then follows that

$$V(n_1, y(n_1, n_0, y')) \geq a(\varepsilon) > b(2\delta_1) \geq b(\delta_1 + \delta_2),$$

and $V(n_0, y') < b(\delta_1)$. Then there exists an index $n_2 \in [n_0, n_1 - 1]$ such that

$$V(n_2, y(n_2, n_0, y')) \leq b(\delta_1 + \delta_2)$$

and (4.11.5)

$$V(n_2 + 1, y(n_2 + 1, n_0, y')) \geq b(\delta_1 + \delta_2).$$

Thus $\|y(n_2 + 1, n_0, y')\| \geq \delta_1 + \delta_2$ from which we get

$$\begin{aligned} \|\tilde{y}(n_2 + 1, n_2, y(n_2, n_0, y')\| &\geq \|y(n_2 + 1, n_0, y')\| \\ &\quad - \|y(n_2 + 1, n_2, y(n_2, n_0, y')) \\ &\quad - \tilde{y}(n_2 + 1, n_2, y(n_2, n_0, y'))\| \\ &\geq \delta_1 + \delta_2 - \delta_2 = \delta_1. \end{aligned}$$

Then from (4.11.5) and (4.11.4), it follows that

$$\begin{aligned} 0 &\leq V(n_2 + 1, y(n_2 + 1, n_0, y')) - V(n_2, y(n_2, n_0, y')) \\ &< -c(\|\tilde{y}(n_2 + 1, n_2, y(n_2, n_0, y')\|) + c(\delta_1) \leq 0, \end{aligned}$$

which is a contradiction. ■

Corollary 4.11.1. *Suppose that the hypothesis of Theorem 4.11.1 is verified and moreover, for $y_n \in B_a$ one has $\|\boldsymbol{R}(n, y_n)\| < g_n \|y_n\|$ with $g_n \to 0$ monotonically. Then the solution of the perturbed equation is uniformly asymptotically stable.*

Proof. From (4.11.4) one has

$$\Delta V(n, y_n) \leq -c(\|\tilde{y}(n + 1, n, y_n)\|) + Mg_n \|y_n\|.$$

Suppose $0 < r < \delta(\delta_0)$, and $r < \|\tilde{y}(n + 1, n, y_n)\| < \delta(\delta_0)$, by the hypothesis on g_n, it can be chosen an $n_1 \in \boldsymbol{N}^+_{n_0}$ such that for $n \geq n_1$ one has $Mg_n \|y_n\| < 2^{-1} c(r)$ and then

$$\Delta V(n, y_n) \leq -\tfrac{1}{2} c(\|\tilde{y}(n + 1, n, y_n).$$

Then apply Theorem 4.8.5. ■

Connected to the problem of total stability is the problem of practical stability of (4.11.2) and (4.11.1). In this case we no longer require that $\boldsymbol{R}(n, 0) \neq 0$ so that (4.11.1) does not have the fixed point at the origin, but it is known that $\|\boldsymbol{R}(n, 0)\|$ is bounded for all n. This kind of stability is very important in numerical analysis, where certain kinds of errors cannot be made arbitrarily small.

Definition 4.11.2. The solution $y = 0$ of (4.11.2) is said to be practically stable, if there exists a neighborhood A of the origin and $\bar{n} \geq n_0$ such that for $n \geq \bar{n}$ the solution $y(n, n_0, y_0)$ of (4.11.2) remains in A.

Theorem 4.11.2. *Consider the equation* (4.11.1) *and suppose that in a set* $D \subset \boldsymbol{R}$ *the following conditions are satisfied*

$$(1)\ \|f(n, y) - f(n, y')\| \leq L\|y - y'\|, \qquad L < 1,$$

$$(2)\ \|\boldsymbol{R}(n, y)\| < \delta.$$

Then the origin is practically stable for (4.11.2).

Proof. Let y_n and $\tilde{y}_n$ be the solution of (4.11.1) and (4.11.2) respectively. Set $m_n = \|y_n - \tilde{y}_n\|$ then by hypothesis $m_{n+1} \leq Lm_n + \delta$ from which it follows

$$\|y_n - \tilde{y}_n\| \leq L^n\|y_0 - \tilde{y}_0\| + \delta \sum_{j=0}^{n-1} L^j = L^n\|y_0 - \tilde{y}_0\| + \frac{\delta}{1 - L}.$$

If $y_0 = \tilde{y}_0$, we see that the distance between the two solutions will never exceed $\dfrac{\delta}{1 - L}$. Thus choosing $n > \bar{n}$ suitably, both of the solutions will remain in the ball $B\left(0, \dfrac{\delta}{1 - L}\right)$ and the proof is complete. ■

The next theorem generalizes the previous result.

Theorem 4.11.3. *Consider the equation* (4.11.1) *and a set* $D \subset \boldsymbol{R}^s$. *Suppose there exists two continuous real valued functions defined on D such that, for all* $x \in D$

$$(1)\ V(x) \geq 0,$$

$$(2)\ \Delta V = V(f(n, x) + \boldsymbol{R}(n, x)) - V(x) \leq w(x) \leq a,$$

for some constant $a \geq 0$. *Let* $S = \{x \in \bar{D} \,|\, w(x) \geq 0\}$, $b = \sup\{V(x) \,|\, x \in S\}$ *and* $A = \{x \in \bar{D} \,|\, V(x) \leq b + a\}$. *Then every solution, which remains D and enters A for* $n = \bar{n}$, *remains in A for* $n \geq \bar{n}$.

Proof. Suppose that $y_{\bar{n}} = y(\bar{n}, n_0, y_0) \in A$, then $V(y_{\bar{n}}) \leq b + a$ and $V(y_{\bar{n}+1}) \leq V(y_{\bar{n}}) + w(y_n)$. If $w(y_{\bar{n}})$ is less than zero, $y_{\bar{n}} \notin S$, then $V(y_{\bar{n}+1}) \leq b + a$ from which it follows that $y_{\bar{n}+1} \in A$. If $y_{\bar{n}} \in S$, then, because $V(y_{\bar{n}}) \leq b$, again it follows that $V(y_{\bar{n}+1}) \leq b + a$. The proof is complete by induction. ■

Corollary 4.11.2. *If* $\delta = \sup\{w(x) \,|\, x \in D - A\} < 0$ *then each solution* $y(n)$ *of* (4.11.1), *which remains in D enters A in a finite number of steps.*

Proof. From $\Delta V \le w(y)$ we get $V(y_n) \le V(y_{n_0}) + \sum_{j=n_0}^{n} w(y_j)) \le V(y_{n_0}) + \delta(n - n_0)$, from which it follows that $V(y(n)) \to -\infty$ as $n \to \infty$, and this is a contradiction because $V(y_n) \ge b + a$ for $y_n \in D - A$. ■

4.12. Problems

4.1. Show that if in the equation $x_{n+1} = a_n x_n$, the product $|\prod_{i=n_0}^{\infty} a_i|$ is bounded, then the zero solution is uniformly stable.

4.2. Consider the following equation, $y_{n+1} = \dfrac{n+1}{2} y_n^2$. Compute, by using a calculator, the two solutions starting from $y_0 = 10^{-21}$, $n = 0$ and $y_0 = -1$, $n_0 = 2\,10^2$. Conclude that the zero solution is not uniformly stable.

4.3. Show that for the system (4.1.8), the origin is the only fixed point that is attractive but not stable.

4.4. Let $y \equiv (x, z)$ be such that $y_{n+1} = \begin{pmatrix} x_n^2 - z_n^2 \\ 2x_n z_n \end{pmatrix}$. Show that the set $D = \{(x, z) | x^2 + z^2 \le 1\}$ is the region of asymptotic stability of the origin. (Hint: Apply Theorem 4.9.3. Take $\phi(y) = x^2 + z^2$ and show that if there is convergence, then $\prod_{j=1}^{\infty} (1 + \phi(y_j)) = \dfrac{1}{1 - x_0^2 - z_0^2}$ from which $V(y_0) = x_0^2 + z_0^2$).

4.5. Consider the linear scalar equation $y_{n+1} = \dfrac{\log(n+2)}{\log(n+3)} y_n$. Find the solutions and show that the zero solution is asymptotically but not exponentially stable. (Hint: Put $z_n = y_n \log(n+2)$).

4.6. Show that the zero solution of the previous problem is not l_p stable for any $p \ge 1$.

4.7. Show that the function $\delta(\varepsilon, n_0)$ in the definition of stability can be taken of class K. (Hint: Take $\delta_1(\varepsilon) = \sup \delta(\varepsilon, n_0)$. The function $\delta_1(\varepsilon)$ is positive and strict increasing. Take $\psi \in K$ such that $\psi(\varepsilon) \le \delta_1(\varepsilon)$. This function can be used instead of δ in the definition.)

4.8. Show that an alternative definition of stability is: The trivial solution is stable if there exists $\phi \in K$ and $n_1 \in N_{n_0}^+$ such that for $n \ge n_1$, $\|y(n, n_0, y_0)\| \le \phi(\|y_0\|, n_0)$. (Hint: Take as ϕ the inverse of ψ defined in the previous problem.)

4.9. Using Theorem 4.8.5 study the behavior of the solutions of the equation $y_{n+1} = y_n^{-2}$.

4.10. Consider the equation $y_n = Ay_{n-1} + b$, where

$$A = \begin{pmatrix} 0 & & & & & & & & \\ & 0 & & & & & & & \\ & & 0 & & & & & -\frac{1}{2} & \\ & & & 0 & & & & & \\ & & & & 0 & & & & \\ & & & & & 0 & & & \\ & -\frac{1}{2} & & & & & 0 & & \\ & & & & & & & 0 & \\ & & & & & & & & 0 \end{pmatrix}_{s\times s}$$

and study the stability of the critical point. The problem arises in economics treating oligopoly systems. The case $s = 2$ was treated by Cournot in finding asymptotic stability.

4.11. Show that the function $g(n, u) = u_n - \omega(u_n)$ in Corollary 4.8.3 is not decreasing and prove that the origin is asymptotically stable for (4.8.6).

4.12. Show that $\Omega(y_0)$, the limit set of $y_0 \in \boldsymbol{R}^s$, is invariant and closed.

4.13. Consider the equation $y_{n+1} = \dfrac{\eta - \frac{1}{2}y_n^2}{1 - y_n}$, $y_0 = \eta$. Determine the values of η for which y_n converges and find the limit point.

4.14. Consider the system

$$x_{n+1} = y_n$$

$$y_{n+1} = x_n + f(x_n),$$

and suppose that $\Delta(x_n f(x_n)) > 0$ for all n. Show, by using Theorem 4.8.7 that the solutions are either unbounded or tend to zero. (Hint: Take $V_n = x_n y_n$.)

4.13. Notes

The definitions in Section 4.1 are modified versions of those given for differential equations, for example, see Corduneanu [30] and Lakshmikantham and Leela [90]. Theorem 4.1.1 is adapted from Gordon [61]. The results of Section 4.2 are adapted from Corduneanu [29]. Most of the material of Section 4.4 has been taken from Halanay [68, 69, 70], see also [46]. The contents of Sections 4.5, 4.7, and 4.8 have been adapted from Halanay [68] and Lakshmikantham and Leela [90]. For Theorem 4.8.4, see Gordon [61]. Theorems 4.8.5 and 4.8.6 are due to Hurt [81] (see also LaSalle [88]). Theorem 4.9.3 is the discrete version of Zubov's theorem and can be found in O'Shea [120] and Ortega [125]. Theorem 4.9.4 is due to Diamond [43, 44]. The converse theorems are adapted from Lakshmikantham and Leela [90] and Halanay [68]. The total stability in the discrete case is treated by Halanay [68] and Ortega [125]. Practical stability is discussed in Hurt [81] and Ortega [125].

CHAPTER 5 Applications to Numerical Analysis

5.0. Introduction

Despite the fact that the two theories have been developed almost independently, the connections between numerical analysis and the theory of difference equations are several. In this chapter, we shall explore some of these connections. In Sections 5.1 to 5.3, iterative methods for solving nonlinear equations are discussed and the importance of employing the theory of difference inequalities is stressed. Sections 5.4 and 5.5 deal with certain classical algorithms from the point of view of the theory of difference equations. Sections 5.6 and 5.7 are devoted to the study of monotone iterative techniques, which offer monotone sequences that converge to multiple solutions of nonlinear equations. This study also includes extension of monotone iterative methods to nonlinear equations with singular linear part as well as applications to numerical analysis. In Section 5.8 we provide problems of interest.

5.1. Iterative Methods

By iterative methods one usually means methods by which one is able to reach a root of a linear or nonlinear equation. Let us consider the problem of solving the algebraic equation of the form

$$F(x) = 0, \tag{5.1.1}$$

where $F: \boldsymbol{R}^s \to \boldsymbol{R}^s$, which is usually transformed into the iterative form:

$$x_{n+1} = f(x_n), \qquad x_{n_0} = x_0. \tag{5.1.2}$$

The function $f(x)$ is called the iterative function and is defined such that the fixed points of f are solutions of (5.1.1). There are many ways to define the iterative function f. The essential criterion is that the root x^* of (5.1.1), which we are interested in, is asymptotically stable for (5.1.2), although it may not be sufficient to ensure rapidity of convergence. This, however, will imply that if x_0 lies in the region of asymptotic stability of x^*, the sequence will converge.

If f is linear, we know that asymptotic stability can be recognized by looking at the eigenvalues of A (the iterative matrix) and indeed this is what is done in the study of linear iterative methods such as Jacobi, Gauss–Seidel, SOR, etc. Moreover, in the linear case, asymptotic stability implies global asymptotic stability. For nonlinear equations the situation becomes more difficult. In this case there are essentially three kinds of results that one can discuss:

A. LOCAL RESULTS. These results ensure asymptotic stability of x^*, but they have nothing to say on the region of asymptotic stability. Usually theorems in this category start with the unpleasant expression "If x_0 is sufficiently near to x^* . . . ."

B. SEMILOCAL RESULTS. The results in this category verify that an auxiliary positive function (usually the norm of the difference of two consecutive terms of the sequence) is decreasing along the sequence itself. The sequence is supposed to (or one proves that it) lie in a closed set $D \subset \boldsymbol{R}^s$. Then one can infer that the auxiliary function has a minimum in D and this minimum is located at x^*. We shall see that this usually requires that x^* is exponentially stable. Connected to these types of results are those requiring the stronger condition of contractivity of f in a closed set $D \subset \boldsymbol{R}^s$. Contractivity implies exponential stability of the fixed point x^*.

C. GLOBAL RESULTS. These results say that the sequence x_n given by (5.1.2) is almost (except, on a set of measure zero) globally convergent.

Of course, results in the class (C) are very few in the nonlinear case. One such result is due to Barna and Smale, which says that for a polynomial with only real root, Newton's method is almost globally convergent, which we shall not present here. We shall, however, discuss in the next two sections some of the most important results in the classes (A) and (B).

5.2. Local Results

Let us begin with the following main result. Here $f: \boldsymbol{R}^s \to \boldsymbol{R}^s$ and x^* satisfies $x^* = f(x^*)$.

Theorem 5.2.1. *Suppose that f is differentiable with Lipshitz derivative, and the spectral radius of $f'(x^*)$ is less than one. Then x^* is asymptotically stable.*

Proof. By the transformation $e_\eta = x_\eta - x^*$, from (5.1.2) we obtain the difference equation

$$e_{n+1} = f(x_n) - f(x^*) = f'(x^*)e_n + \int_0^1 \{f'(x^* + s(x_n - x^*)) - f'(x^*)\}\, ds\, e_n \tag{5.2.1}$$
$$= f'(x^*)e_n + g(e_n)$$

where $\lim_{n\to\infty} \frac{\|g(e_n)\|}{\|e_n\|} = 0$, (see problem 5.2). The equation (5.2.1) is in the form required by Corollary 4.7.2, which assures the exponential asymptotic stability of x^*.

Note that the exponential asymptotic stability does not imply in general that $\|x_1 - x^*\| < \|x_0 - x^*\|$, unless (see section 4.2) $\alpha\eta < 1$, where the constant α depends on the norm used.

The result established in Chapter 4 can be used to prove the next theorem on the convergence of nonstationary iterative methods defined by

$$x_{n+1} = q(n, x_n), \tag{5.2.2}$$

where $q: N^+_{n_0} \times D \subset R^s \to R^s$, $x_0 \in D$. It is obvious that nonstationary iterative methods are, in our terminology, nonautonomous difference equations. ■

Theorem 5.2.2. *Consider the difference equation*

$$y_{n+1} = f(y_n), \qquad y_{n_0} = x_0, \tag{5.2.3}$$

where $f: D \to R^s$, and f is locally Lipshitzian. Suppose that $y^* \in D$ is an asymptotic stable solution for (5.2.3) and that the solutions $y(n, n_0, x_0)$ of (5.2.3) and $x(n, n_0, x_0)$ of (5.2.2) are in D for $n \in N^+_{n_0}$. Assume further that

$$\|q(n, x) - f(x)\| \le L_n \|x - y^*\|, \tag{5.2.4}$$

where $L_n \to 0$ for $n \to \infty$. Then the solution $x(n, n_0, x_0) \to y^*$, when x_0 is suitably chosen.

Proof. Rewrite (5.2.2) as

$$x_{n+1} = f(x_n) + R(n, x_n) \tag{5.2.5}$$

and consider (5.2.5) as the perturbed equation relative to (5.2.3) with

$$R(n, x_n) = q(n, x_n) - f(x_n) \tag{5.2.6}$$

as the perturbation, which tends to zero as $x_n \to y^*$. Then apply Corollary 4.11.1. ■

Other results of local type can be obtained by using the comparison principle given in section 1.6. For example:

Theorem 5.2.3. *Let $f: D \subset \boldsymbol{R}^s \to \boldsymbol{R}^s$ continuous, $g: \boldsymbol{R}^+ \to \boldsymbol{R}^+$. Suppose that*

(1) *there exists $x^* \in \operatorname{Int} D$ such that for all $x \in B(x^*, \delta) \subset D$, $\delta > 0$,*

$$\|f(x) - x^*\| \leq \|x - x^*\| + g(\|x - x^*\|); \tag{5.2.7}$$

(2) *$G(u) = u + g(u)$ is nondecreasing with respect to u and $g(0) = 0$;*
(3) *the trivial solution of*

$$u_{n+1} = G(u_n),\ u_0 = \|x_0 - x^*\|,$$

is asymptotically stable, with u_0 in the domain of asymptotic stability, and $u_i \leq u_0$ for all i. Then x^* is an asymptotically stable fixed point for the equation

$$x_{n+1} = f(x_n). \tag{5.2.8}$$

Proof. Let $x_0 \in B(x^*, \delta)$. Then by Theorem 1.6.1, we have $\|x_n - x^*\| \leq u_n$ and, because u_n is decreasing to zero, all $x_n \in B(x^*, \delta)$ and the sequence will tend to zero. ■

The next two theorems are applications of LaSalle's Theorem 4.8.6. Here we study the convergence of secant and Newton methods on the real line. These two methods are widely used in numerical analysis to find roots of equations and their generalizations are almost countless. On the real line, these two methods are defined by

$$z_{n+2} = z_{n+1} - \frac{(z_{n+1} - z_n)}{f(z_{n+1}) - f(z_n)} f(z_{n+1}), \tag{5.2.9}$$

and

$$z_{n+1} = z_n - f(z_n)^{-1} f(z_n), \tag{5.2.10}$$

respectively. If $f(z_k) \neq f(z_{k+1})$, the difference equation (5.2.9) is well defined.

Suppose that f is differentiable and that α is the simple root of $f(z) = 0$. Using the transformation

$$e_n = z_n - \alpha \tag{5.2.11}$$

and the mean value theorem

$$f(\alpha + e) = f'(\alpha) + g(\alpha, e)e^2, \tag{5.2.12}$$

we write the new difference equation corresponding to (5.2.9) having the fixed point at the origin in the form

$$e_{n+2} = M(\alpha, e_n, e_{n+1})e_n e_{n+1}, \tag{5.2.13}$$

where

$$M(\alpha, e_n, e_{n+1}) = \frac{g(\alpha, e_{n+1})e_{n+1} - g(\alpha, e_n)e_n}{f(\alpha + e_{n+1}) - f(\alpha + e_n)}. \tag{5.2.14}$$

If α is a simple root and $g(\alpha, e)$ is continuous and bounded, then $M(\alpha, e_k, e_{k-1})$ is continuous and bounded if e_k, e_{k-1} are small enough. The second order difference equation (5.2.13) can be transformed into a first order system by introducing the variables $x_n = e_n$ and $y_n = e_{n+1}$. Then the system is

$$x_{n+1} = y_n,$$

$$y_{n+1} = M(\alpha, x_n, y_n)x_n y_n.$$

By using the Lyapunov function $V(x, y) = |x|^q + |y|^q$, $q \geq 1$, one obtains

$$\Delta V(x_n, y_n) = -(1 - |M(\alpha, x_n, y_n)y_n|^q)|x_n|^q,$$

which is negative if $W(x, y) \equiv (1 - |M(\alpha, x, y)y|) > 0$. Let

$$D(\eta) = \{(x, y) \,|\, (|x|^q + |y|^q)^{1/q} < \eta\}.$$

Since $y = 0$ implies $W(x, y) = 1 > 0$, there exists some $\eta > 0$ such that $W(x, y) \geq 0$ in $D(\eta)$.

If the initial values z_0 and z_1 are such that (x_0, y_0) and $(x_1, y_1) \in D(\eta)$, then (x_n, y_n) will remain in D (see Corollary 4.8.5) and will converge to the maximum invariant set contained in the set $E = \{(x, y) \in \bar{D}(\eta) \,|\, x = 0\}$. The only invariant set with $x_1 = 0$ is the origin, so we obtain $(x_n, y_n) \to (0, 0)$ for $n \to \infty$.

Corollary 4.8.5 can be used to study the convergence of (5.2.10). Because there is no more difficulty involved we shall consider the case where (5.2.10) is defined in $\boldsymbol{R}^s$. The root α, which we are interested in, is simple. The change of variable similar to (5.2.11) allows us to write the method in the form

$$e_{n+1} = M_1(e_n)[M_2(e_n)e_n - f(\alpha + e_n)]$$

where

$$M_1(e_n) = \left[\frac{\partial f(\alpha + e_n)}{\partial x}\right]^{-1} \quad \text{and} \quad M_2(e_n) = \left[\frac{\partial f(\alpha + e_n)}{\partial x}\right].$$

It can be shown that for $\|e_n\| < \eta$,

$$\|M_1(e_n)[M_2(e_n)e_n - f(\alpha + e_n)]\| \leq k(\eta)\|e_n^2\|.$$

Then, taking $V(e) = \|e\|$, one can obtain

$$\Delta V(e) \leq -(1 - k(\eta)\|e\|)\|e\| \equiv -W(\|e\|),$$

showing that $\Delta V \leq 0$ for $k(\eta)\|e_n\| \leq 1$. Using Corollary 4.8.5, we get a region of convergence $D(\eta_0) = \{x \mid \|x - \alpha\| < \eta_0\}$ with $\eta_0 = \min(\eta, 1/k(\eta))$. The parameter η can be chosen to maximize η_0 and consequently the region of convergence.

5.3. Semilocal Results

Results of semilocal type are obtained either by requiring contractivity or some of its generalizations on a compact $D \subset \boldsymbol{R}^s$ or by requiring the weaker condition that some auxiliary function decreases over a sequence. We shall consider some cases of the latter type. We begin discussing the simplest of such results in some detail to bring out clearly the arguments used.

Theorem 5.3.1. *Suppose that all x_n obtained by* (5.1.2) *lie in a compact $D_0 \subset R^s$ and that*

$$\|x_{n+2} - x_{n+1}\| \leq \alpha \|x_{n+1} - x_n\|, \tag{5.3.1}$$

with $\alpha < 1$. Then the sequence converges to $x^* \in D_0$.

Proof. Setting

$$y_n = \|x_{n+1} - x_n\|, \tag{5.3.2}$$

we have, because of (5.3.1),

$$y_{n+1} \leq \alpha y_n.$$

We can now apply Theorem 1.6.1 to obtain $y_n \leq u_n$, where u_n is the solution of

$$u_{n+1} = \alpha u_n \tag{5.3.4}$$

provided that $y_0 \leq u_0$.

Now the equation (5.3.4) has the solution $u_n = 0$ exponentially asymptotically stable, which means (see Theorem 4.1.1) that $\sum_{j=1}^{\infty} u_j$ is bounded. In fact from (5.3.4) we get $\sum_{j=0}^{\infty} u_j = \dfrac{1}{1-\alpha} u_0$. It then follows that $\|x_{j+1} - x_j\| \leq u_j$, $j = 1, 2, \ldots$, if we choose $u_0 = \|x_1 - x_0\|$. Moreover, for $p \geq 1$, we obtain

$$\begin{aligned} \|x_{n+p} - x_n\| &\leq \|x_{n+p} - x_{n+p-1} + x_{n+p-1} - x_{n-p-2} + \ldots + x_{n+1} - x_n\| \\ &\leq \sum_{j=0}^{p-1} \|x_{n+j+1} - x_{n+j}\| \leq \sum_{j=0}^{p-1} u_{n+j}. \end{aligned}$$

Because the right-hand side can be made less than an arbitrary positive quantity when n is chosen properly, this inequality shows that x_n is a Cauchy sequence and thus must converge to x^* in D_0. From this result we can also find the distance between x_0 and x^*. In fact, from

$$\|x_n - x_0\| \le \sum_{j=0}^{n-1} \|x_{j+1} - x_j\| \le \sum_{j=0}^{n-1} u_j,$$

it follows that

$$\|x^* - x_0\| \le \sum_{j=0}^{\infty} u_j = \frac{1}{1-\alpha} u_0. \tag{5.3.5}$$

Using this property one can avoid the a priori assumption that all x_n lie in D_0. Also, if we assume that f is continuous, then $f(x^*) = x^*$. Furthermore, if we assume more than (5.3.1), namely, $x, y \in D$, $\|f(x) - f(y)\| \le k < 1$, then the solution x^* is unique in D_0. ■

Using similar arguments, one can also prove the following result.

Theorem 5.3.2. *Suppose that x_0, x_1 lies in D_0 and $\overline{B(x_0, \sum_{j=0}^{\infty} u_j)} \subset D_0$. Then the sequence converges to $x^* \in \overline{B(x_0, \sum_{j=0}^{\infty} u_j)}$.*

The next theorems are on the convergence of Newton's method defined by

$$x_{n+1} = x_n - F'(x_n)^{-1} F(x_n), \tag{5.3.6}$$

where F is the same function as in (5.1.1). The main result was first proved by Kantorovich and is usually called Newton–Kantorovich theorem.

Theorem 5.3.3. *Assume that, for $x_0, x, y \in D$, where D is a convex subset of $\mathbf{R}^s$,*

(1) *F is differentiable in D;*

(2) $\|F'(x) - F'(y)\| \le \gamma \|x - y\|$;

(3) $\|F'(x_0)^{-1}\| \le \beta, \qquad \|F'(x_0)^{-1} F(x_0)\| = \eta$;

(4) $\beta\gamma\eta < \frac{1}{2}$; *and*

(5) $\overline{B\left(x_0, \dfrac{1}{\beta\gamma}\right)} \subset D.$

Then the sequence defined by (5.3.6) is well defined and converges to a root x^* of $F(x) = 0$ contained in $\overline{B(x_0, 1/\beta\gamma)}$.

Proof. For every $x \in B(x_0, 1/\beta\gamma)$ we have

$$\|F'(x) - F'(x_0)\| \le \gamma \|x - x_0\| < \beta^{-1}.$$

A bound of $\|F'(x)^{-1}\|$ can then be obtained by using the Banach lemma (see Corollary A5.2), which yields

$$\|F'(x)^{-1}\| \leq \frac{\|F'(x_0)^{-1}\|}{1 - \|F'(x_0)^{-1}\| \, \|F'(x) - F'(x_0)\|} \leq \frac{\beta}{1 - \beta\gamma\|x - x_0\|}.$$

From this estimate and from hypothesis (4), it follows that for $x \in B(x_0, 1/\beta\gamma)$, $F'(x)$ is nonsingular and the Newton's iterative function is defined. Moreover,

$$\|x_{n+2} - x_{n+1}\| = \|F'(x_{n+1})^{-1}F(x_{n+1})\| \leq \frac{\beta\|F(x_{n+1})\|}{1 - \beta\gamma\|x_{n+1} - x_0\|}, \tag{5.3.7}$$

and (see problem 5.2)

$$\begin{aligned} \|F(x_{n+1})\| &= \|F(x_{n+1}) - F(x_n) - F'(x_n)(x_{n+1} - x_n)\| \\ &\leq \tfrac{1}{2}\gamma\|x_{n+1} - x_n\|^2. \end{aligned} \tag{5.3.8}$$

Substituting for $F(x_{n+1})$ in (5.3.7), we then find

$$\|x_{n+2} - x_{n+1}\| \leq \frac{\beta\gamma}{2} \frac{\|x_{n+1} - x_n\|^2}{1 - \beta\gamma\|x_{n+1} - x_0\|}.$$

Suppose for a moment that the sequence x_n lies in $B(x_0, 1/\beta\gamma)$, then we can apply Theorem 4.5.4 with $g(u_n, \sum_{j=1}^{n} u_j, \sum_{j=1}^{n-1} u_j) = \dfrac{\beta\gamma u_n^2}{2(1 - \beta\gamma \sum_{j=0}^{n} u_j)}$. The comparison equation is therefore

$$u_{n+1} = \frac{\beta\gamma u_n^2}{2\left(1 - \beta\gamma \sum_{j=0}^{n} u_j\right)}, \qquad u_0 = \frac{1}{2}(\beta\gamma)^{-1}, \tag{5.3.9}$$

whose solution is $u_n = (\beta\gamma)^{-1}2^{-n-1}$. The origin is thus exponentially stable for (5.3.9). All x_n lie in $B(x_0, 1/\beta\gamma)$ because $\|x_n - x_0\| < (\beta\gamma)^{-1}\sum_{j=1}^{n-1} 2^{-j-1} < \dfrac{1}{\beta\gamma}$. We may apply the same arguments used before. In fact, $\|x_{n+p} - x_n\| \leq \sum_{j=0}^{p-1} \|x_{n+j+1} - x_{n+j}\| \leq \sum_{j=0}^{p-1} u_{n+j} = \dfrac{(\beta\gamma)^{-1}}{2} \times \sum_{j=0}^{p-1} 2^{-n-j} = (\beta\gamma)^{-1}2^{-n}(1 - 2^{-p})$, which shows that $x_n \to x^*$ and also gives the error estimate $\|x^* - x_n\| \leq (\beta\gamma)^{-1}2^{-n}$. The proof that x^* is a root of $F(x)$ follows from $F(x_n) = F'(x_n)(x_{n+1} - x_n) = [F'(x_0) - (F'(x_n) - F'(x_0))](x_{n+1} - x_n)$ and then $\|F(x_n)\| \leq [\|F'(x_0)\| + \gamma\|x_n - x_0\|]\|x_{n+1} - x_n\| \leq [\|F'(x_0)\| + \gamma(\beta\gamma)^{-1}]\|x_{n+1} - x_n\|$. Taking the limit as $n \to \infty$ one has, by the continuity of F, $F(x^*) = 0$.

Hence by Theorem 4.5.4, the conclusion follows. ■

The conclusion of Theorem 5.3.3 has been easily obtained because the solution of (5.3.9) can be given in the closed form. More refined estimates can be obtained if we assume in (5.3.9) that $u_0 < \frac{1}{2}(\beta\gamma)^{-1}$. In this case, the solution of (5.3.9) is more difficult to compute but it can be done.

In fact, let us define the new sequence z_n by $\Delta z_{n-1} = \beta\gamma u_n$, $z_{-1} = 0$. Then equation (5.3.9) is transformed into $z_{n+1} = z_n + \dfrac{1}{2}\dfrac{(z_n - z_{n-1})^2}{1 - z_n}$, which possesses the first integral, that is, the sequence $\{z_n\}$ satisfies the first order equation $z_{n+1} = \dfrac{z_0 - \frac{1}{2}z_n^2}{1 - z_n}$, where $z_0 = \beta\gamma u_0$. (See (1.5.12).) The solution of this equation is $z_n = 1 - (1 - 2z_0)^{1/2} \coth k2^n$, (see problem 1.27) where k is a constant depending on the initial condition. (See problem 5.8 for its determination and for a more refined bound.) This solution exists if $z_0 \leq \frac{1}{2}$ and then $u_0 \leq \frac{1}{2}(\beta\gamma)^{-1}$.

The case of equality has already been examined. Suppose that $u_1 < \frac{1}{2}(\beta\gamma)^{-1}$. Then it follows that $\lim_{n\to\infty} z_n = 1 - (1 - 2z_0)^{1/2} \equiv z^*$ and since for large n, $\coth k2^n \simeq 1 + 2\,e^{-k2^{n+1}}$, we obtain $z_n - z^* \simeq 2(1 - 2z_0)^{1/2}\,e^{-k2^{n+1}}$. Consequently, we get $\lim_{n\to\infty} \dfrac{z_n - z^*}{(z_{n-1} - z^*)^2} = 1$, which shows that the order of convergence of the sequence z_n, to the limit point, is quadratic.

From the comparison principle it follows that $\|x_{n+1} - x_n\| < \dfrac{1}{\beta\gamma}(z_n - z_{n-1})$ and hence $\|x_n - x^*\| < \dfrac{2}{\beta\gamma}(1 - 2z_0)^{1/2}\,e^{-k2^{n+1}}$. Also for $n > 0$, $\|x_n - x_0\| \leq \sum_{j=0}^{n-1} u_j \leq \dfrac{1}{\beta\gamma} z_{n-1} \leq \dfrac{1}{\beta\gamma} z^*$ because of the increasing behavior of z_n. ■

Summing up these considerations we see that we have proved a variant of Theorem 5.3.3, which we state below.

Theorem 5.3.3(a). *Assume that the hypotheses* (1), (2), (3) *of Theorem* 5.3.3 *hold. Suppose further that* (4) $\beta\gamma\eta < \frac{1}{2}$, *and* (5) $B\left(x_0, \dfrac{1}{\beta\gamma} z^*\right) \subset D$ *are satisfied. Then the sequence x_n is well defined and converges quadratically to x^* contained in* $B\left(x_0, \dfrac{1}{\beta\gamma} z^*\right)$.

Newton's method is invariant under affine transformation, that is, it is invariant under the transformation $G = AF$ where A is an invertible matrix, whereas the hypotheses in the Newton–Kantarovich theorem are not. In fact, one has $G^{-1}G = F^{-1}A^{-1}AF = F^{-1}F$, while for example, the hypothesis (3) is not invariant under the same transformation. The following variant of Theorem 5.3.3 takes care of this situation.

Theorem 5.3.3(b). *Assume that the hypotheses* (1) *of Theorem* 5.3.3 *holds and suppose that*

(2) $\|F'(x_0)^{-1}F(x_0)\| \leq a$;
(3) $\|F'(x_0)^{-1}(F'(y) - F'(x))\| \leq w\|x - y\|$;
(4) $aw \leq \frac{1}{2}$;
(5) $B(x_0, w^{-1}z^*) \in D$, *where* $z^* = 1 - (1 - 2wa)^{1/2} \leq 1$.

Then the sequence x_n *remains in* $B(x_0, w^{-1}z^*)$ *and converges quadratically to a root* x^* *of* $F(x)$.

Proof. We have

$$\begin{aligned} x_{n+2} - x_{n+1} &= -F'(x_{n+1})^{-1}F(x_{n+1}) \\ &= -F'(x_{n+1})^{-1}[F(x_n) + F(x_{n+1}) - F(x_n)] \\ &= -F'(x_{n+1})^{-1}\Big\{F(x_n) + F'(x_n)(x_{n+1} - x_n) \\ &\qquad + \int_0^1 [F'(x_n + t(x_{n+1} - x_n)) - F'(x_n)]\,dt(x_{n+1} - x_n)\Big\} \\ &= -F'(x_{n+1})^{-1}F'(x_0)\Big\{\int_0^1 F'(x_0)^{-1}[F'(x_n + t(x_{n+1} - x_n)) \\ &\qquad - F'(x_n)]\,dt(x_{n+1} - x_n)\Big\}, \end{aligned}$$

from which it follows that

$$\|x_{n+2} - x_{n+1}\| \leq \|F'(x_{n+1})^{-1}F'(x_0)\| \cdot \frac{w}{2}\|x_{n+1} - x_n\|^2.$$

Since $F'(x_0)^{-1}F'(x_{n+1}) = F'(x_0)^{-1}F'(x_{n+1}) - F'(x_0)^{-1}F'(x_0) + F'(x_0)^{-1}F(x_0)$, using hypothesis (3), we get $\|F'(x_0)^{-1}(F'(x_{n+1}) - F'(x_0))\| < w\|x_{n+1} - x_0\| < z^* \leq 1$, if $x_{n+1} \in B(x_0, w^{-1}z^*)$. Then, by the Banach lemma it follows that $F'(x_{n+1})$ is invertible and $\|F'(x_{n+1})^{-1}F'(x_0)\| \leq \dfrac{1}{1 - w\|x_{n+1} - x_0\|}$. Finally, we arrive at

$$\|x_{n+2} - x_{n+1}\| \leq \frac{w}{2}\,\frac{\|x_{n+1} - x_n\|^2}{1 - w\|x_{n+1} - x_0\|},$$

which is the same as before with $\beta\gamma$ replaced by w. All the previous results then apply, and we are done. ■

The next result also shows the familiar quadratic decay of the errors in the Newton method, but it requires more assumptions on the inverse of the Jacobian.

Theorem 5.3.4. *Assume that $F: D_0 \subset \boldsymbol{R}^s \to \boldsymbol{R}^s$ and $D_0 \subset \boldsymbol{R}^s$ is a convex set. Moreover, suppose that*

(1) *for $x \in D_0$, F is differentiable,*
(2) *for $x \in D_0$, $F'(x)$ is invertible and $\|F'(x)^{-1}\| \le \beta$,*
(3) *for $x, y \in D_0$, $\|F'(x) - F'(y)\| \le \gamma \|x - y\|$,*
(4) *for $x_0 \in D_0$, $\|F'(x_0)^{-1}F(x_0)\| \le \eta$, and*
(5) $\alpha = \frac{1}{2}\beta\gamma\eta < 1$.

Then the Newton iterates remain in $\overline{B(x_0, r_0)}$, where $r_0 = \eta \sum_{j=0}^{\infty} \alpha^{2^j - 1}$ and converge to a solution x^* of $F(x) = 0$.

Proof. From

$$\|x_{n+2} - x_{n+1}\| = \|F'(x_{n+1})^{-1}F(x_{n+1})\| \le \beta \|F(x_{n+1})\|$$

and from the estimate of $\|F(x_{n+1})\|$ obtained in the previous theorem, we get

$$\|x_{n+2} - x_{n+1}\| \le \frac{\beta\gamma}{2} \|x_{n+1} - x_n\|^2. \tag{5.3.10}$$

The comparison equation is therefore

$$u_{n+1} = \frac{\beta\gamma}{2} u_n^2, \qquad u_0 = \eta, \tag{5.3.11}$$

whose solution is $u_n = \eta\alpha^{2^n - 1}$. Now applying comparison Theorem 1.6.1 we get (see problem 5.4)

$$\|x_n - x_{n+1}\| \le u_n = \eta\alpha^{2^n - 1}. \tag{5.3.12}$$

Since $\alpha < 1$, the trivial solution of (5.3.11) is exponentially stable and hence it follows that x_n is a Cauchy sequence. Furthermore, we also have $\|x_n - x_0\| \le \sum_{j=0}^{\infty} u_j = r_0$, which means that all the iterates remain in $\overline{B(x_0, r_0)}$ and converge to x^*. ■

Corollary 5.1.1. *The error $\|x^* - x_n\|$ satisfies the inequality*

$$\|x^* - x_n\| \le \varepsilon_n \|x_n - x_{n-1}\|^2, \qquad \text{for } n = 1, 2, \ldots. \tag{5.3.13}$$

As an application of Theorem 1.6.7, let us consider the two-step iterative method

$$x_{k+1} = G(x_k, x_{k-1}).$$

Theorem 5.3.5. *Suppose that $G: D \times D \subset \boldsymbol{R}^s \times \boldsymbol{R}^s \to \boldsymbol{R}^s$ and on some closed set $D_0 \subset \boldsymbol{R}^s$, the following inequality holds true*

$$\|G(x, y) - G(y, z)\| \leq \alpha \|x - y\| + \beta \|y - z\|,$$

where $\alpha + \beta < 1$. Furthermore, let there exist $x_0, x_1 \in D$ such that $x_k \in D_0$ for $k \geq 1$. Then the iteratives converge to the unique fixed point of $\hat{G}(x) = G(x, x)$.

Proof. Setting $y_k = \|x_{k+1} - x_k\|$, we obtain $y_{k+1} \leq \alpha y_k + \beta y_{k-1}$. By Theorem 1.6.7, we then get $y_k \leq u_k$, where u_k is the solution of $u_{k+1} = \alpha u_k + \beta u_{k-1}$, which is $u_k = c_0 z_1^k + c_1 z_2^k$, where z_1 and z_2 are the roots of $z^2 - \alpha z - \beta = 0$. Because of the assumption on α and β, the two roots z_1 and z_2 are less than 1 in modulus, which means that the zero solution of the comparison equation is exponentially stable. Using similar arguments as in the previous theorems, we can conclude that the sequence x_k is a Cauchy sequence. Since

$$\begin{aligned}\|\hat{G}(x^*) - x_{k+1}\| &\leq \|G(x^*, x^*) - G(x^*, x_{k-1})\| \\ &\quad + \|G(x^*, x_{k-1}) - G(x_k, x_{k-1})\| \\ &\leq \beta \|x^* - x_{k-1}\| + \alpha \|x^* - x_k\|,\end{aligned}$$

the convergence of x_k shows that $x_{k+1} \to \hat{G}(x^*)$. The uniqueness of x^* now follows very easily and the proof is complete. ■

The effect of perturbation on the iterative methods are of two different kinds. If the perturbations are small and tend to zero as n tends to ∞, then the theorems on total stability will ensure that the fixed points continue to have asymptotic stability. In numerical analysis, however, perturbations may remain bounded for all n (roundoff errors, for example). In this case, a more convenient concept is the practical stability. If an iterative procedure is practically stable, the sequence of iterates will not tend to the solution x^* but to a ball $B(x^*, \delta)$ surrounding x^*. Inside B, the solution may oscillate but what is important is that it never leaves B. Of course it would be nice to have B as small as possible.

Let us consider for example the iterative method

$$x_{n+1} = f(x_n), \qquad x_{n_0} = x_0. \tag{5.3.14}$$

We assume that for $x, y \in D_0 \subset \boldsymbol{R}^s$,

$$\|f(x) - f(y)\| \leq \alpha \|x - y\| \tag{5.3.15}$$

with $\alpha < 1$. Suppose that the errors in the computations perturb (5.3.14) by a bounded perturbation $\boldsymbol{R}_n$, that is

$$\tilde{x}_{n+1} = f(\tilde{x}_n) + R_n, \qquad \tilde{x}_{n_0} = x_0, \tag{5.3.16}$$

with x_n, $\tilde{x}_n \in D_0$ for all n. Then Theorem 4.11.2 can be applied. The conclusion is that the difference of the two solutions $\|x_n - \tilde{x}_n\|$ will not exceed $\boldsymbol{R}/(1-\alpha)$, where $\boldsymbol{R} = \sup \boldsymbol{R}_n$. The condition that all the balls $B\left(x_n, \dfrac{\boldsymbol{R}}{1-\alpha}\right)$ must be contained in D_0 may be very restrictive. In fact if α is close to one, it follows that these balls are very large and the method is considered a bad one.

5.4. *Unstable Problems: Miller's, Olver's and Clenshaw's Algorithms*

The situation that we shall analyze in this section and in the next is in a certain sense the opposite of that treated in the previous ones. In fact, there the limit point was important (and indeed in the theory of iterative processes the "solution" is the limit point), where the intermediate values z_n are considered an unavoidable noise. Here the limit point will not be important (generally will not exist) and the important part ("the solution") will be a finite subset of the sequence itself.

These problems are very typical in that part of numerical analysis which concerns the study of special functions, orthogonal polynomials, quadrature formulas, and numerical methods for ordinary differential equations.

Consider the homogeneous scalar linear equation (see (2.1.5)) of order k.

$$Ly_n = 0. \tag{5.4.1}$$

It has k linearly independent solutions $f_1(n), f_2(n), \ldots, f_k(n)$. Suppose that the solution $y_n = 0$ $(n \geq 0)$ is unstable, and

$$\lim_{n\to\infty} \frac{f_1(n)}{f_i(n)} = 0, \qquad i = 2, 3, \ldots, k. \tag{5.4.2}$$

When this happens, the solution $f_1(n)$ is said to be minimal. The problem is how to find f_1 or a multiple of it. Even if we know the exact initial condition that generates the minimal solution, small errors in the calculations will, as usual, lead us to solve a perturbed equation $Ly_n = \varepsilon_n$, whose solution (see (2.1.11)) will contain all the $f_i(n)$ and this will destroy the process (see for example [52], [175]), because the $f_i(n)$ $(i \geq 2)$ grow faster than $f_1(n)$. The same problem appears, of course, in the nonhomogeneous case, where it is not excluded a priori that there exists a bounded solution even if all the $f_i(n)$ are unbounded. One is often interested in following this solution.

Consider, for example, the following first order equation

$$y_{n+1} - (n+1)y_n = -1, \qquad y_0 = e. \tag{5.4.3}$$

The solution of the homogeneous part is $y_n = en!$, while the complete solution is $y_n = n!\left(e - \sum_{j=0}^{n} \frac{1}{j!}\right) \equiv \sum_{s=1}^{\infty} \frac{1}{(n+s)^{(s)}}$, which is bounded for $n \to \infty$. If one tries to follow this solution, a small error in the initial data e (always existing because it has infinite digits), will be amplified by the factor $n!$.

A simple idea is to find the way to look for the solution of the problem backwards (for $n \to -\infty$), because in this case the origin becomes asymptotically stable for the homogeneous equation and the errors will be damped. This idea indeed works, as we will show in a simple case, which is the original problem to which it was applied (see [52] for references).

Consider

$$Ly_n \equiv a_n y_{n+1} + b_n y_n + c_n y_{n-1}, \tag{5.4.6}$$

where a_n, $c_n \neq 0$ for all n, and the nonhomogeneous problem

$$Ly_n = g_n. \tag{5.4.7}$$

Suppose that two linearly independent solutions ϕ_n and ψ_n of the homogeneous equation are such that ϕ_n is minimal, that is

$$\lim_{n\to\infty} \frac{\phi_n}{\psi_n} = 0. \tag{5.4.8}$$

Of course all solutions that are multiple of ϕ_n are minimal and vice versa. That means that only one appropriate initial condition is needed to determine the minimal solution we are interested in. The general solution of the nonhomogeneous problem is then

$$y_n = c_1\phi_n + c_2\psi_n + \bar{y}_n, \tag{5.4.9}$$

where $\bar{y}_n$ is a particular solution that will be supposed minimal too; that is,

$$\lim_{n\to\infty} \frac{\bar{y}_n}{\psi_n} = 0, \tag{5.4.10}$$

and $\bar{y}_0 = 0$.

We are interested in the solution

$$y_n = \frac{y_0}{\phi_0}\phi_n + \bar{y}_n. \tag{5.4.11}$$

A class of methods are based on the following result.

Theorem 5.4.1. *Consider the boundary value problem*

$$Ly^{(N)} = g_n, \qquad y_0^{(N)} = y_0, \qquad y_N^{(N)} = 0, \tag{5.4.12}$$

where N is a positive integer, and suppose that conditions (5.4.8) *and* (5.4.10) *are verified. Then the problem* (5.4.12) *has a solution and moreover, for fixed n, $y_n^{(N)} \to y_n$ as $\mathbf{N} \to \infty$.*

Proof. Since y_n is a particular solution of the nonhomogeneous equation, any other solution of the same equation will be

$$y_n^{(N)} = c_1^{(N)}\phi_n + c_2^{(N)}\psi_n + y_n. \tag{5.4.13}$$

The boundary conditions give for $c_1^{(N)}$ and $c_2^{(N)}$ the values

$$c_1^{(N)} = \frac{\dfrac{y_N}{\psi_N}}{\dfrac{\phi_0}{\psi_0} - \dfrac{\phi_N}{\psi_N}}; \qquad c_2^{(N)} = -\frac{\phi_0}{\psi_0}c_1^{(N)}.$$

It follows that $c_1^{(N)}$ and $c_2^{(N)}$ tend to zero as $\mathbf{N} \to \infty$ and the claim follows.

Note that the condition $y_N^{(N)} = 0$ can be replaced by $y_N^{(N)} = k$ where k is any fixed value (for example an approximation of y_N if available).

In the homogeneous case $y_n^{(N)}$ can be obtained starting from $\tilde{y}_N^{(N)} = 0$, $\tilde{y}_{N-1}^{(N)} = 1$, and then using the difference equation backwards

$$\tilde{y}_{n-1}^{(N)} = -\frac{a_n}{c_n}\tilde{y}_{n+1} - \frac{b_n}{c_n}\tilde{y}_n$$

obtaining a value $\tilde{y}_0^{(N)}$ different from y_0. Multiplying then the sequence by the scale factor $y_0/\tilde{y}_0^{(N)}$, one obtains $y_n^{(N)}$. This is the Miller's algorithm. A disadvantage of this algorithm is that one does not know a priori which value of $\mathbf{N}$ must be used. A modified version of it, which works in the nonhomogeneous case is also obtained by considering that the boundary value problem (5.4.9) is equivalent to the system of equations

$$Ay^{(N)} = b, \tag{5.4.14}$$

where

$$A = \begin{pmatrix} b_1 & a_1 & 0 & \dots & 0 \\ c_1 & & & & \vdots \\ 0 & & & & 0 \\ \vdots & & & & a_{N-1} \\ 0 & \dots & 0 & c_{N-1} & b_{N-1} \end{pmatrix}$$

and

$$y^{(N)} = (y_1^{(N)} \dots y_{N-1}^{(N)})^T,$$
$$b = (g_1 - c_1 y_0, \dots, g_{N-1})^T.$$

Some clever method to solve this system can be used to avoid the growth of the errors (see [23], [24]). By solving the system in an appropriate way, it is also possible to determine the best value of N (Olver's algorithm). ■

Similar in motivation to the previous algorithms and indeed related to them is the following, known as Clenshaw's algorithm. Here the problem is to evaluate the sum $\sum_{n=0}^{N} b_n u_n$, where b_n is a given sequence and u_n is supposed to satisfy a kth linear difference equation $Lu_n = 0$. The algorithm uses the result stated in Theorem 2.1.8, that is, one considers the transposed equation $L^T y_n = b_n$, $y_{N+j} = 0$ $(j = 1, 2, \ldots, k)$, obtaining

$$\sum_{n=0}^{N} b_n u_n = \sum_{j=0}^{k-1} y_j \sum_{i=0}^{j} p_i(j-k) u_{j-i}.$$

Suppose, for example, we want to compute $f_N = \sum_{n=0}^{N} b_n T_n(z)$, where $T_n(z)$ are Chebyshev polynomials satisfying the equation $T_{n+1} - 2zT_n + T_n = 0$. The transpose equation is

$$y_n - 2zy_{n+1} + y_{n+2} = b_n, \qquad n = N, N-1, \ldots, 0$$

$$y_{N+1} = y_{N+2} = 0,$$

which can be solved recursively obtaining y_1 and y_0. From the quoted result it then follows $f_N = y_0 + y_1[T_1(z) - 2zT_0(z)] = y_0 - zy_1$. It is interesting to note that the sum can be obtained without the knowledge of the polynomials $T_n(z)$ except for $T_0(z)$ and $T_1(z)$ and also the reduction of the operations involved.

5.5. Unstable Problems: the "SWEEP" Method

The topic discussed in this section is very similar to the previous one, but historically it has different sources. It arises typically in that part of numerical analysis that concerns the approximation of solutions of ODE and PDE. We shall consider, for simplicity, the autonomous case. The more general case is not, in principle, more complicated.

Let us consider

$$y_{n+1} - 2\delta y_n + y_{n-1} = g_n, \qquad y(0) = \alpha, \qquad y_k = \beta, \qquad (5.5.1)$$

with $\delta > 1$, $k > 0$. This problem can be solved in many different ways. It can be solved rewriting it as a system of $k-1$ equations similar to (5.4.14). In order to control the growth of errors it is better to treat the problem as a difference equation.

The characteristic polynomial of (5.5.1) has two positive roots, one less than one and the other greater than one. This means that the origin is not asymptotically stable for the homogeneous part and this, as usual, gives troubles because the errors are amplified.

One way to avoid this difficulty is to transform the linear second order equation (5.5.1) into nonlinear first order equations (see problem 5.12).

$$x_{n+1} = \frac{1}{2\delta - x_n}, \tag{5.5.2}$$

and

$$z_{n+1} = \frac{-g_n + z_n}{2\delta - x_n}. \tag{5.5.3}$$

Furthermore, one verifies that

$$y_{n-1} = x_n y_n + z_n. \tag{5.5.4}$$

The solution is then obtained by computing recursively the sequences $\{x_n\}$, $\{z_n\}$ using (5.5.2) and (5.5.3) with $x_1 = 0$ and $z_1 = \alpha$ up to $n = k - 1$ and then computing $\{y_n\}$ using (5.5.4) from $n = k$ to $n = 1$ with $y_k = \beta$. The advantage is that (5.5.2) has a limit point asymptotically stable and the sequence x_n converges monotonically to this point, while the sequences (5.5.3) and (5.5.4) remain bounded.

Theorem 5.5.1. *If* $x_0 = 0$, $\delta > 1$, *the sequence* $\{x_n\}$ *converges monotonically to the first root* ρ *of the characteristic polynomial of* (5.5.1) *and, moreover,*

$$0 \le x_n < \rho.$$

Proof. The critical points of (5.5.2) are the roots of the equation $x^2 - 2\delta x + 1 = 0$, which is the characteristic polynomial of (5.5.1). Let ρ be the root less than one. Changing the variable $h_n = \rho(\rho - x_n)$, equation (5.5.2) becomes

$$h_{n+1} = \frac{\rho^2}{1 + h_n} h_n. \tag{5.5.5}$$

One shows at once (for example using the Lyapunov function $V_n = h_n^2$) that the half line $h > 0$ is contained in the asymptotic stability region of the origin, and moreover $0 < h_n \le \rho^2$, from which it follows that

$$-\rho \le x_n - \rho < 0. \quad \blacksquare \tag{5.5.6}$$

The next theorem is not really necessary in this case, but it could be useful in the more general case.

Theorem 5.5.2. *In the hypothesis of Theorem* 5.5.1 *the solution* $x_n = \rho$ *is practically stable for* (5.5.2).

Proof. The conditions of Theorem 4.11.2 are satisfied for $x_n < \rho$. ■

The last result ensures that for equation (5.5.2) the errors will not be amplified.

Let us consider the equations (5.5.3) and (5.5.4).

Theorem 5.5.3. *In the hypothesis of Theorem* 5.5.1, *the sequence* $\{z_n\}$ *remains bounded, even if the two equations are perturbed with a bounded perturbation.*

Proof. From $\rho^2 - 2\delta\rho + 1 = 0$, it follows that $2\sigma - \rho = 1/\rho$ and (5.5.3) can be written as $z_{n+1} = \rho \dfrac{-g_n + z_n}{1 - \rho(x_n - \rho)}$ and from (5.5.6) $|z_{n+1}| \leq \rho(|g_n| + |z_n|)$. Using Corollary 1.6.1, it follows that $|z_n| \leq \rho^n \alpha + \sum_{s=0}^{n-1} \rho^{n-s+1}|g_s|$ from which we see that if $|g_s|$ is bounded z_n is bounded. Clearly if we add to g_n a bounded perturbation τ_n (any sort of errors), the statement remains true. ■

Similar arguments can be applied to (5.5.4) since it follows that $|y_{n-1}| \leq \rho|y_n| + |z_n|$. The method just described is known as the "SWEEP" method (see [28]).

5.6. Monotone Iterative Methods

In this section, we shall develop a general theory for a broad class of monotone iterations. This class of iterations includes Newton's method as well as a family of methods, which are called Newton–Gauss–Seidel processes that are obtained by using the Gauss-Seidel iteration on the linear systems of Newton's method.

As before, we are interested in finding the solutions of

$$0 = f(u), \tag{5.6.1}$$

where $f \in C[\mathbf{R}^n; \mathbf{R}^n]$. Let us first split the system (5.6.1) as

$$0 = f_i(u_i, [u]_{p_i}, [u]_{q_i}), \tag{5.6.2}$$

where, for each i, $1 \leq i \leq n$, $p_i + q_i = n - 1$ and $u = (u_i, [u]_{p_i}, [u]_{q_i})$.

Let $v, w \in \mathbf{R}^n$ be such that $v \leq w$. Then v, w are said to be coupled quasi lower and upper solutions of (5.6.1) if

$$0 \leq f_i(v_i, [v]_{p_i}, [w]_{q_i}),$$

$$0 \geq f_i(w_i, [w]_{p_i}, [v]_{q_i}).$$

Coupled quasi extremal solutions of (5.6.1) can be defined easily. Vectorial inequalities mean the same inequalities between the components of the vectors.

We are now in a position to prove the following result.

Theorem 5.6.1. *Assume that $f \in C[\boldsymbol{R}^n, \boldsymbol{R}^n]$ and possesses a mixed quasimonotone property, that is, for each i, $f_i(u_i, [u]_{p_i}, [u]_{q_i})$ is nondecreasing in $[u]_{p_i}$ and nonincreasing in $[u]_{q_i}$. Suppose further v, w are coupled quasi lower and upper solutions of* (5.6.1) *and*

$$f_i(u_i, [u]_{p_i}, [u]_{q_i}) - f_i(\bar{u}_i, [u]_{p_i}, [u]_{q_i}) \geq -M_i(u_i - \bar{u}_i)$$

whenever $v \leq \bar{u} \leq u \leq w$ and $M_i > 0$. Then there exist monotone sequences $\{v_n\}$, $\{w_n\}$ such that $v_n \to \rho$, $w_n \to r$ as $n \to \infty$ and ρ, r are coupled minimal and maximal solutions of (5.6.1) *such that $v \leq \rho \leq u \leq r \leq w$ for any solution u of* (5.6.1).

Proof. For any $\eta_1, \eta_2 \in [v, w] = [u \in \boldsymbol{R}^n : v \leq u \leq w]$. Consider the system

$$u_i = M_i^{-1} f_i(\eta_{1i}, [\eta_1]_{p_i}, [\eta_2]_{q_i}) + \eta_{1i}, \qquad i = 1, 2, \ldots, n.$$

Clearly u can be uniquely defined, given $\eta_1, \eta_2 \in [v, w]$. Therefore, we can define a mapping A such that $A[\eta_1, \eta_2] = u$. It is easy to show that A satisfies the properties:

(i) $0 \leq A[v, w]$, $0 \geq A[w, v]$;
(ii) A is mixed monotone on $[v, w]$.

Then the sequences $\{v_n\}$, $\{w_n\}$ with $v_0 = v$, $w_0 = w$ can be defined as follows

$$v_n = A[v_{n-1}, w_{n-1}], \qquad w_n = A[w_{n-1}, v_{n-1}].$$

Furthermore, it is clear that $\{v_n\}$, $\{w_n\}$ are monotone sequences such that $v \leq v_n \leq w_n \leq w$ and consequently $\lim_{n\to\infty} v_n = \rho$, $\lim_{n\to\infty} w_n = r$ exist and satisfy the relations

$$0 = f_i(\rho_i, [\rho]_{p_i}, [r]_{q_i}), \qquad 0 = f_i(r_i, [r]_{p_i}, [\rho]_{q_i}).$$

By induction, it is also easy to show that if (u_1, u_2) is a coupled quasi solution of (5.6.1) such that $v \leq u_1$, $u_2 \leq w$, then $v_n \leq u_1$, $u_2 \leq w_n$ and consequently (ρ, r) are coupled quasi extremal solutions. Since any solution u of (5.6.1) is a coupled quasi solution, the conclusion of the theorem follows and the proof is complete. ■

If f does not possess the mixed quasi monotone property, we need a different set of assumptions to generate monotone sequences that converge to extremal solutions. This is the content of the following results.

Theorem 5.6.2. *Assume that*

(i) *there exist* $v, w \in \boldsymbol{R}^n$ *with* $v \leq w$ *such that* $0 \leq f(v)$, $0 \geq f(w)$;
(ii) *there is a* $n \times n$ *matrix* M *such that*

$$f(y) - f(x) \geq -M(y)(y - x),$$

whenever $v \leq x \leq y \leq w$.

Then the sequence $\{w_n\}$, *given by*

$$w_{n+1} = w_n + B_n f(w_n), \tag{5.6.3}$$

where B_n *is any nonnegative subinverse of* $M(w_n)$, *is well defined, and* $\{w_n\}$ *is monotone nonincreasing such that* $\lim_{n\to\infty} w_n = r$. *If there exists a nonsingular matrix* $B \geq 0$ *such that* $\liminf_{n\to\infty} B_n \geq B$, *then* r *is a maximal solution of* (5.6.1).

Proof. Set $v = v_0$ and $w = w_0$. From $B_0 > 0$ and $f(w_0) \leq 0$ it follows that $w_1 \leq w_0$. Using the fact that B_0 is a subinverse of $M(w_0)$, we find for any $u \in [v, w]$,

$$\begin{aligned} u + B_0 f(u) &= w_1 - (w_0 - u) - B_0[f(w_0) - f(u)] \\ &\leq w_1 - [I - B_0 M(w_0)](w_0 - u) \leq w_1. \end{aligned}$$

Hence, in particular, $v_0 \leq v_0 + B_0 f(v_0) \leq w_1$. Similarly, we obtain

$$f(w_1) \leq f(w_0) + M(w_0)(w_0 - w_1) = [I - M(w_0)B_0]f(w_0) \leq 0.$$

Proceeding similarly we see by induction that

$$w_{n-1} \geq w_n \geq v_0, \qquad f(w_n) \geq 0, \quad n = 1, 2, \ldots. \tag{5.6.4}$$

Consequently, as a monotone nonincreasing sequence that is bounded below, $\{w_n\}$ has a limit $r \geq v_0$.

If u is any solution of (5.6.1) such that $u \in [v, w]$, then $u = u + B_0 f(u) \leq w_1$, then by induction $u \leq w_n$ for all n. Hence $u \leq r$. Finally, the continuity of f and the fact $\liminf_n B_n \geq B$, where $B \geq 0$ is nonsingular, (5.6.4) yields

$$\begin{aligned} 0 &= \liminf_{n\to\infty} [w_{n+1} - w_n = B_n f(w_n)] \\ &= -(\liminf_{n\to\infty} B_n) f(r) \geq -Bf(r) \geq 0, \end{aligned}$$

which implies $f(r) = 0$ completing the proof. ■

By following the argument of Theorem 5.6.2, one can prove the next corollary.

Corollary 5.6.1. *Let the assumptions of Theorem* 5.6.2 *hold. Suppose that* $M(y)$ *is monotone nonincreasing in* y. *Then the sequence* $\{v_n\}$ *with* $v_0 = v$, *given by*

$$v_{n+1} = v_n + B_n f(v_n)$$

is well defined, monotone nondecreasing such that $\lim_{n\to\infty} v_n = \rho$, *and is the minimal solution of* (5.6.1).

The case of most interest is when $\rho = r$, because then the sequences $\{v_n\}$, $\{w_n\}$ constitute lower and upper bounds for the unique solution of (5.6.1). The following uniqueness result is of interest.

Theorem 5.6.3. *Suppose that*

$$f(y) - f(x) \leq \mathbf{N}(x)(y - x),$$

where $\mathbf{N}(x)$ *is nonsingular matrix and* $\mathbf{N}(x)^{-1} \geq 0$. *Then if* (5.6.1) *has either maximal or minimal solution in* $[v, w]$, *then there are no other solutions in* $[v, w]$.

Proof. Suppose $r \in [v, w]$ is the maximal solution of (5.6.1) and $u \in [v, r]$ is any other solution of (5.6.1). Then

$$0 - f(r) - f(u) \leq \mathbf{N}(u)(r - u).$$

Since $\mathbf{N}(u)^{-1} \geq 0$, it follows that $r \geq u$ and hence $r = u$. A similar proof holds if the minimal solution exists. The proof is complete. ■

5.7. Monotone Iterative Methods (Continued)

Consider the problem of finding solutions of

$$Ax = f(x), \tag{5.7.1}$$

where A is an $n \times n$ matrix and $f \in C[\mathbf{R}^n, \mathbf{R}^n]$, which arises as finite difference approximation to nonlinear differential equations. If A is nonsingular, writing $F(x) = f(x) - Ax = 0$, one can study existence of multiple solutions by employing the method of upper and lower solutions and the monotone iterative technique, described in Section 5.6. In this section we extend such existence results to equation (5.7.1) when A is singular. For convenience let us split the system (5.7.1) and write in the form

$$(Au)_i = f_i(u_i, [u]_{p_i}, [u]_{q_i}), \tag{5.7.2}$$

where for each i, $1 \le i \le n$, $u = (u_i, [u]_{p_i}, [u]_{q_i})$ with $p_i + q_i = n - 1$ and $(Au)_i$ represents the ith component of the vector Au. As before, a function $f \in C[\boldsymbol{R}^n, \boldsymbol{R}^n]$ is said to be mixed quasi-monotone if, for each i, f_i is monotone nondecreasing relative to $[u]_{p_i}$ components and monotone nonincreasing with respect to $[u]_{q_i}$ components.

Let $v, w \in \boldsymbol{R}^n$ be such that $v \le w$. Then v, w are said to be coupled quasi lower and upper solutions of (5.7.1) if

$$(Av)_i \le f_i(v_i, [v]_{p_i}, [w]_{q_i}), \qquad (Aw)_i \ge f_i(w_i, [w]_{p_i}, [v]_{q_i}).$$

Coupled quasi-extremal solutions and solutions can be defined with equality holding.

We are now in a position to prove the following result.

Theorem 5.7.1. *Assume that*

(i) *$f \in C[\boldsymbol{R}^n, \boldsymbol{R}^n]$ and f is mixed quasi-monotone*;

(ii) *v, w are coupled quasi lower and upper solutions of* (5.7.1);

(iii) *$f_i(u_i, [u]_{p_i}, [u]_{q_i}) - f_i(\bar{u}_i, [u]_{p_i}, [u]_{q_i}) \ge -M_i(u_i - \bar{u}_i)$ whenever $v \le \bar{u} \le u \le w$ and $M_i > 0$, for each i*;

(iv) *A is an $n \times n$ singular matrix such that $A + M = \boldsymbol{C}$ is nonsingular where M is a diagonal matrix with $M_i > 0$ and $\boldsymbol{C}^{-1} \ge 0$.*

Then there exist monotone sequences $\{v_n\}$, $\{w_n\}$ such that $v_n \to \rho$, $w_n \to r$ as $n \to \infty$ and ρ, r are coupled quasi-extremal solutions of (1.1) *such that if u is any solution of* (1.1), *then $v \le \rho \le u \le r \le w$.*

Proof. For any $\eta, \mu \in [v, w] = [x \in \boldsymbol{R}^n : v \le x \le w]$, consider the linear system

$$\boldsymbol{C}u = F(\eta, \mu), \tag{5.7.3}$$

where $\boldsymbol{C} = A + M$ and for each i, $F_i(\eta, \mu) = f_i(\eta_k, [\eta]_{p_i}, [\mu]_{q_i}) + M_i\eta_i$. Clearly u can be uniquely defined given $\eta, \mu \in [v, w]$ since the matrix $\boldsymbol{C}$ is nonsingular. Furthermore, we have $Av \le F(v, w)$, $Aw \ge F(w, v)$, and F is mixed monotone. Consequently we can define a mapping T such that $T[\eta, \mu] = u$ and show easily, using the fact that $\boldsymbol{C}^{-1} \ge 0$, that

(a) $v \le T[v, w]$, $w \ge T[w, v]$;

(b) T is mixed monotone on $[v, w]$.

Then the sequences $\{v_n\}$, $\{w_n\}$ with $v_0 = v$, $w_0 = w$ can be defined as follows:

$$v_{n+1} = T[v_n, w_n], \qquad w_{n+1} = T[w_n, v_n].$$

Furthermore, it is evident from the properties (a) and (b) that $\{v_n\}$, $\{w_n\}$ are monotone such that $v_0 \le v_1 \le \ldots \le v_n \le w_n \le \ldots \le w_1 \le w_0$. Consequently, $\lim_{n\to\infty} v_n = \rho$, $\lim_{n\to\infty} w_n = r$ exist and satisfy the relations

$$(A\rho)_i = f_i(\rho_i, [\rho]_{p_i}, [r]_{q_i}), \qquad (Ar)_i = f_i(r_i, [r]_{p_i}, [\rho]_{q_i}). \tag{5.7.4}$$

By induction, it is also easy to prove that if (u_1, u_2) is a coupled quasisolution of (5.7.1) such that $v \le u_1$, $u_2 \le w$, then $v_n \le u_1$, $u_2 \le w_n$ for all n and hence (ρ, r) are coupled quasi-extremal solutions of (5.7.1). Since any solution u of (5.7.1) is a coupled quasi-solution, the conclusion of the theorem follows and the proof is complete. ■

The case of most interest is when $\rho = r$, because then the sequences $\{v_n\}$, $\{w_n\}$ constitute lower and upper bounds for the unique solution of (5.7.1). The following uniqueness result is thus of interest.

Corollary 5.7.1. *If, in addition to the hypotheses of Theorem* 5.7.1, *we assume for each i, that*

$$f_i(x_i, [x]_{p_i}, [y]_{q_i}) - f_i(y_i, [y]_{p_i}, [x]_{q_i}) \le [B(x - y)]_i, \tag{5.7.5}$$

where B is an $n \times n$ matrix such that $(A - B)$ is nonsingular and $(A - B)^{-1} \ge 0$. Then $u = \rho = r$ is the unique solution of (5.7.1) *such that $v \le u \le w$.*

Proof. Since $\rho \le r$, it is enough to prove that $\rho \ge r$. We have by (5.7.5)

$$[A(r - \rho)]_i \le f_i(r_i, [r]_{p_i}, [\rho]_{q_i}) - f_i(\rho_i, [\rho]_{p_i}, [r]_{q_i}) \le [B(r - \rho)]_i.$$

Consequently, $(A - B)(r - \rho) \le 0$, which implies $r \le \rho$. ■

Remark 5.7.1. If $q_i = 0$ for each i, then Theorem 5.7.1 shows that r, ρ are maximal and minimal solutions of 5.7.1 and Corollary 5.7.1 gives the unique solution.

If f does not possess the mixed quasi monotone property, we need a different set of assumptions to generate monotone sequences that converge to a solution. This is the content of the next result.

Theorem 5.7.2. *Assume that* (iv) *of Theorem* 5.7.1 *holds. Further, suppose that $v, w \in \boldsymbol{R}^n$ such that $v \le w$,*

$$Av \le f(v) - B(w - v), \qquad Aw \ge f(w) + B(w - v), \tag{5.7.6}$$

$$-B(x - y) \le f(x) - f(y) \le B(x - y) \tag{5.7.7}$$

whenever $v \le y \le x \le w$, *B being an* $n \times n$ *matrix of nonnegative elements. Then there exist monotone sequences* $\{v_n\}$, $\{w_n\}$ *that converge to a solution* u *of* (5.7.1) *such that*

$$v \le v_1 \le \ldots \le v_n \le u \le w_n \le \ldots \le w_1 \le w,$$

provided that $(A - B)$ *is a nonsingular matrix.*

Proof. We define

$$F(y, z) = \tfrac{1}{2}[f(y) + f(z) + B(y - z)]. \tag{5.7.8}$$

It is easy to see that $F(y, z)$ is mixed monotone and

$$-B(z - \bar{z}) \le F(y, z) - F(\bar{y}, \bar{z}) \le B(y - \bar{y}) \tag{5.7.9}$$

whenever $z, \bar{z}, y, \bar{y} \in [v, w]$ and $\bar{z} \le z$, $\bar{y} \le y$. In particular, we have

$$F(y, z) - F(z, y) = B(y - z). \tag{5.7.10}$$

From (5.7.8) we obtain $-B(w - v) \le F(v, w) - f(v)$, which yields $Av \le f(v) - B(w - v) \le F(v, w)$. Similarly $Aw \ge F(w, v)$. Finally, it follows from (5.7.8) that $F(x, x) = f(x)$.

Consider now for any $\eta, \mu \in [v, w]$, the linear system given by $Cu = G(\eta, \mu)$ where $C = A + M$, M being the diagonal matrix with $M_i > 0$ and for each i, $G_i(\eta, \mu) = F_i(\eta, \mu) + M_i\eta_i$. Proceeding as in proof of Theorem 5.7.1, we arrive at $A\rho = F(\rho, r)$, $Ar = F(r, \rho)$. Using (5.7.9), we see that $A(r - \rho) = F(r, \rho) - F(\rho, r) = B(r - \rho)$ and this implies $u = r = \rho$ is a solution of (5.7.1). The proof is complete. ■

Corollary 5.7.2. *If in addition to the hypotheses of Theorem* 5.7.2 *we also have* $f(x) - f(y) \le C(y)(x - y)$ *for* $x, y \in [v, w]$, *where* $C(y)$ *is an* $n \times n$ *matrix such that* $[A - C(y)]$ *is nonsingular and* $[A - C(y)]^{-1} \ge 0$, *then* u *is the unique solution of* (5.7.1).

Proof. If $\bar{u}$ is another solution of 5.7.1, we get

$$A(u - \bar{u}) = f(u) - f(\bar{u}) \le C(\bar{u})(u - \bar{u}),$$

which yields $u \le \bar{u}$. Similarly, $\bar{u} \le u$ and hence u is the unique solution of (5.7.1). ■

Equations of the form (5.7.1) arise as finite difference approximations to nonlinear partial differential equations as well as problems at resonance where the matrix is usually singular, irreducible, M-matrix. For such matrices the following result is known.

Theorem 5.7.3. *Let A be an $n \times n$ singular, irreducible, M-matrix. Then* (i) *A has rank $(n-1)$;* (ii) *there exists a vector $u > 0$ such that $Au = 0$;* (iii) *$Av \geq 0$ implies $Av = 0$; and* (iv) *for any nonnegative diagonal matrix D, $(A + D)$ is an M-matrix and $(A + D)^{-1}$ is a nonsingular M-matrix, if $d_{ii} > 0$ for some i, $1 \leq i \leq n$.*

As a consequence of Theorem 5.7.3, if we suppose that A in (5.7.1) is $n \times n$ singular, irreducible, M-matrix, then $(A + M)$ is a nonsingular M-matrix. Therefore assumption (iv) of Theorem 5.7.1 holds since nonsingular M-matrix has the property that its inverse exists and is greater than or equal to zero. Furthermore, in some applications to partial differential equations, the function f in (5.7.1) is of special type, namely $f(u) = (f_1(u_1), f_2(u_2), \ldots, f_n(u_n))$. We can find in this special case lower and upper solutions such that assumption (ii) of Theorem 5.7.1 holds whenever we have

$$\limsup_{|u_i| \to \infty} f_i(u_i) \operatorname{Sig}(u_i) < 0 \quad \text{for each } i.$$

In fact, by Theorem 5.7.3, there exists a $\xi > 0$ such that $\operatorname{Ker} A = \operatorname{Span}(\xi)$ and hence we can choose a $\lambda > 0$ so large that

$$f(\lambda\xi) \leq 0 \quad \text{and} \quad f(-\lambda\xi) \geq 0.$$

Letting $v = -\lambda\xi$, $w = \lambda\xi$, we see that $v \leq w$, $Av \leq f(v)$ and $Aw \geq f(w)$. Assumption (iii) is simply a one-sided Lypschitz condition and in assumption (5.7.5), B is a diagonal matrix.

5.8. Problems

5.1. Show that $\rho(G'(x^*)) = 0$ where $G(x) = x - (F'(x))^{-1}F(x)$ and x^* is a simple root of $F(x) = 0$.

5.2. Show that if $F: \boldsymbol{R}^s \to \boldsymbol{R}$ has a Lypshitz derivative then

$$\|F(x) - F(y) - F'(x)(y - x)\| \leq \tfrac{1}{2}\gamma\|x - y\|^2.$$

5.3. Show that if in the Newton–Kantarovich theorem one supposes $\|F'(x)^{-1}\|$ bounded for all $x \in \boldsymbol{R}^s$, then $\|x_n - x^*\| \leq c\|x_{n-1} - x^*\|^2$.

5.4. Show that in the hypothesis of Theorem 5.3.4, the error satisfies the inequalities (5.3.12) and (5.3.13).

5.5. Suppose that $x_n \in \boldsymbol{R}^s$, $n \geq 0$ and $\|\Delta x_n\| \leq \dfrac{\Delta t_n}{\Delta_{n-1}}\|\Delta x_{n-1}\|$, where t_n is a positive converging sequence with $t_0 = 0$ and $\lim\limits_{n\to\infty} t_n = t^*$. Show that x_n converges to a limit x^*.

5.6. As in the previous exercise suppose that $\|\Delta x_n\| \leq \frac{\Delta t_n}{(\Delta t_{n-1})^\gamma}(\|\Delta x_{n-1}\|)^\gamma$ with $\frac{\|\Delta x_n\|}{\Delta t_n} \leq 1$.

5.7. Show that the solution of $z_{n+1} = \frac{z_0 - \frac{1}{2}z_n^2}{1 - z_n}$ is given by $z_n = 1 - (1 - 2z_0)^{1/2}\,\text{cotgh}(k2^n)$, with k appropriately chosen.

5.8. Find the constant k in the previous problem and deduce that for Newton's method one has:

$$\|x^* - x_n\| \leq \frac{2}{\beta\gamma}(1 - 2z_0)^{1/2}\frac{\theta^{2^n}}{1 - \theta^{2^n}},$$

where

$$\theta = \frac{1 - z_0 - (1 - 2z_0)^{1/2}}{1 - z_0 + (1 - 2z_0)^{1/2}}.$$

5.9. Obtain the result of Theorem 5.3.3 by considering the first integral with $z_0 = \frac{1}{2}$. (Hint: in this case the first order equation becomes $z_{n+1} = \frac{1}{2}(1 + z_n)$.)

5.10. Consider the iterative method

$$x_{k+1} = \mu[\lambda\hat{G}(x_n) + (1 - \lambda)G(x_k, x_{k-1})] + (1 - \mu)x_k,$$

where λ, $\mu \in [0, 1]$ and $\hat{G}$, G are defined as in Theorem 5.3.5. Show that μ and λ can be chosen such that the convergence becomes faster.

5.11. Solve equation (5.5.1) directly and show the growth of the errors.

5.12. Obtain the relations (5.5.2), (5.5.3) and (5.5.4).

5.13. Show that the sweep method is equivalent to the Gaussian elimination of the problem $Ay = g$ when the problem 5.5.1 is stated in vector form, see (5.4.16).

5.9. Notes

The discussions of Sections 5.1 to 5.3 have been introduced only to show examples of application of difference equations to numerical analysis. More details can be found in the excellent books by Ortega and Rheinboldt [127] and Ostrowsky [132]. Theorem 5.3.3 is a simplified version of the original one, see [129], [127], while the estimates given before 5.3.4 (b) seem to be new. For other estimates of the error bounds for the Newton's method and Newton-like methods, see also [41], [109], [111], [179], [144]. For the effect of errors in the iterative methods, see Urabe [168]. A very large source of material on the problem discussed in Section 5.4 can be found in Gautschi's review paper [52]. Theorem 5.4.1 has been taken from Olver [123]. A detailed analysis, in a more general setting, of the Miller's algorithm is given in Zahar [180]. Applications to numerical methods for ODE's can be found in Cash's book [23], while a large exposition of applications as well as theoretical results are in Wimp's book [56], see also Mattheji [104]. The "sweep" method can be found in the Godunov and Rayabenki's book [59], where it is also presented for differential equations. An improvement of the Olver's algorithm can be found in [169]. An application of the sweep method to the problem of Section 5.4 can be found in Trigiante-Sivasundaram [167]. The contents of Section 5.6 are adapted from [89, 126, 127], while the material of Section 5.7 is taken from [91], see also [85, 87, 89].

CHAPTER 6 Numerical Methods for Differential Equations

6.0. Introduction

Numerical methods for differential equations is one of the very rich fields of application of the theory of difference equations where the concepts of stability play a prominent role. Because of the successful use in recent years of computers to solve difficult problems arising in applications such as stiff equations, the connection between the two areas has become more important. Furthermore, in a fundamental work Dahlquist did emphasize the importance of stability theory in the study of numerical methods of differential equations. We shall consider in this chapter some of the most relevant applications. In Section 6.1 we discuss linear multistep methods again and show that the problem can be reduced to the study of total or practical stability when the roundoff errors are taken into account. In Section 6.2 we deal with the case of a finite interval where we shall find a different form of Theorem 2.7.3. Section 6.3 considers the situation when the interval is infinite restricting to the linear case. The nonlinear case when the nonlinearity is of monotone type is investigated in Section 6.4, while in Section 6.5, we show how one can utilize nonlinear variation of constants formula for evaluting global error. Sections 6.6 and 6.7 are devoted to the extension of previous results to partial differential equations via the method of lines, which leads to the consideration of the spectrum of a family of matrices in order to obtain the right stability conditions. The problems given in Section 6.8 complete the picture.

6.1. *Linear Multistep Methods*

We have already seen that a linear multistep method (LM), which approximates the solution of the differential equation

$$y' = f(t, y), \qquad y(t_0) = y_0, \tag{6.1.1}$$

is defined by

$$\rho(E)z_n - h\sigma(E)f(t_n, z_n) = 0, \tag{6.1.2}$$

where $\rho(E)$ and $\sigma(E)$ are defined in section 2.7.

For convenience we shall suppose that f is defined on $\mathbb{R} \times \mathbb{R}$ and it is continuously differentiable. The general case can be treated similarly but for some notational difficulties. The values of the solution $y(t)$ on the knots $t_j = t_0 + jh$ satisfy the difference equation

$$\rho(E)y(t_n) - h\sigma(E)f(t_n, y(t_n)) = \tau_n, \tag{6.1.3}$$

where τ_n is the local truncation error. The global error

$$l_n = y(t_n) - z_n \tag{6.1.4}$$

satisfies then the difference equation

$$\rho(E)l_n - h\sigma(E)[f(t_n, z_n + l_n) - f(t_n, z_n)] = \tau_n. \tag{6.1.5}$$

In practice, equation (6.1.2) is solved on the computer and instead of (6.1.2) one really solves the perturbed equation

$$\rho(E)z_n - h\sigma(E)f(t, z_n) = \varepsilon_n, \tag{6.1.6}$$

where ε_n represent the roundoff errors. The equation for the global error then becomes

$$\rho(E)l_n - h\sigma(E)[f(t_n, z_n + l_n) - f(t_n, z_n)] = w_n, \tag{6.1.7}$$

where

$$\omega_n = \tau_n - \varepsilon_n. \tag{6.1.8}$$

It is worth noting the different nature of the two kinds of errors. If the method is consistent τ_n depend on some power of h, which is greater than one, while ε_n is independent on such a parameter (it depends on the machine precision). For simplicity, we assume that ε_n is bounded. (We note that ε_n can grow as a polynomial in n, see problem 1.14). The equation (6.1.7) can be considered as the perturbation of the equation

$$\rho(E)l_n - h\sigma(E)[f(t_n, z_n + l_n) - f(t_n, z_n)] = 0, \tag{6.1.9}$$

which has the trivial solution $l_n = 0$.

The problem is then a problem of total stability or of practical stability if the roundoff errors ε_n are taken in account. One needs to study equation

(6.1.9) and verify whether the zero solution is uniformly asymptotically stable and then apply, for example, a theorem similar to 4.11.1 in order to get total stability. This has been done indirectly in the linear case. Recently some efforts have been devoted to certain nonlinear cases. The techniques used to solve the problem are different. One method is to use the Z-transform (with arguments similar to those used in the proof of Theorem 2.7.3 (See for example [117]) and the other is to transform the difference equation of order k to a first order system in $\boldsymbol{R}^n$. We shall follow the latter approach.

Let us put

$$f(t_n, z_n + l_n) - f(t_n, z_n) = c_n l_n. \tag{6.1.10}$$

Equation (6.1.9) can be written (we take $\alpha_k = 1$) as

$$l_{n+k} - h\beta_k c_{n+k} l_{n+k} + \sum_{j=0}^{k-1} (\alpha_j - h\beta_j c_{n+j}) l_{n+j} = w_n,$$

or if we suppose that $1 - h\beta_k c_{n+k} \neq 0$ for all values of indices, as

$$\begin{aligned} l_{n+k} &= -\sum_{j=0}^{k-1} \frac{\alpha_j - h\beta_j c_{n+j}}{1 - h\beta_k c_{n+k}} l_{n+j} + \frac{w_n}{1 - h\beta_k c_{n+k}} \\ &= -\sum_{j=0}^{k-1} \alpha_j l_{n+j} - \sum_{j=0}^{k-1} \left[\frac{\alpha_j - h\beta_j c_{n+j}}{1 - h\beta_k c_{n+k}} - \alpha_j\right] l_{n+j} + \frac{w_n}{1 - h\beta_k c_{n+k}} \\ &= -\sum_{j=0}^{k-1} \alpha_j l_{n+j} + h \sum_{j=0}^{k-1} b_j^{(n)} l_{n+j} + \frac{w_n}{1 - h\beta_k c_{n+k}}, \end{aligned}$$

where

$$b_j^{(n)} = \frac{\beta_j c_{n+j} - \beta_k \alpha_j c_{n+k}}{1 - h\beta_k c_{n+k}}.$$

By introducing the k-dimensional vectors

$$\begin{aligned} E_n &= (l_m, l_{n+1}, \ldots, l_{n+k-1})^T, \\ W_n &= \left(0, 0, \ldots, \frac{W_n}{1 - h\beta_k c_{n+k}}\right)^T \\ b^{(n)} &= b^{(n)} = (b_0^{(n)}, b_1^{(n)}, \ldots, b_{k-1}^{(n)})^T; \qquad \phi_k = (0, 0, \ldots 1)^T, \end{aligned} \tag{6.1.12}$$

and the matrices

$$\mathrm{A} = \begin{pmatrix} 0 & 1 & 0 & \ldots & 0 \\ 0 & & & & \vdots \\ & & & & 0 \\ \vdots & & & & 1 \\ 0 & & & 0 & \\ -\alpha_0 & -\alpha_1 & & \ldots & -\alpha_{k-1} \end{pmatrix} \tag{6.1.13}$$

$$B_n = \begin{pmatrix} 0 & \dots & 0 \\ & \dots & \\ & \dots & \\ 0 & \dots & 0 \\ b_0^{(n)} & b_1^{(n)} & b_{k-1}^{(n)} \end{pmatrix} = \Phi_k b^T, \tag{6.1.14}$$

equation (6.1.7) reduces to the form

$$E_{n+1} = (A + hB_n)E_n + W_n. \tag{6.1.15}$$

The problem is now to study total stability for the zero solution of this equation, which depends crucially on the nonlinearity of f contained in the matrix B_n. Historically this problem has been studied under some simplifying hypothesis that we shall summarize as follows: finite interval, infinite interval with f linear, infinite interval with f of monotone type.

6.2. *Finite Interval*

If the interval of integration $(t_0, t_0 + T)$ is bounded, one takes $h = \frac{T}{N}$ and $n = 0, 1, \dots, N$. This implies that h becomes smaller and smaller as N increases and the quantitiy hB_nE_n, like the term W_n, can be considered as small perturbation of the equation

$$E_{n+1} = AE_n. \tag{6.2.1}$$

The eigenvalues of the matrix A are roots of the characteristic polynomial $\rho(z)$ and the requirement of the stability of the null solution of (6.2.1) is equivalent to 0-stability. We cannot assume asymptotic stability of the zero solution and apply theorem 4.11.1 because one of the conditions of consistency impose $\rho(1) = 0$, which means that at least one of the eigenvalues of A is on the boundary of the unit disk in the complex plane. The analysis of the stability of the zero solution of (6.1.15) needs to be done directly. This is not difficult if we use (4.6.2). In fact

$$E_n = A^n E_0 + \sum_{j=0}^{n-1} A^{n-j-1}(W_j + hB_jE_j). \tag{6.2.3}$$

Taking the norms, we get

$$\|E_N\| \leq \|A^N\| \, \|E_0\| + \sum_{j=0}^{N-1} \|A^{N-j-1}\|(\|W_j\| + h\|B_j\| \, \|E_j\|). \tag{6.2.4}$$

Now suppose that the zero solution of (6.2.1) is stable (that is the same to demand that the method is 0-stable). This means that the spectral radius of A is one and moreover the eigenvalues on the boundary of the unit circle are simple. It follows then (see Theorem 12.4) that the powers of A are bounded. Let us take for simplicity that $\|A^j\| = 1, j = 1, 2, \ldots, N$, (see also Theorem A5.2). The inequality (6.2.3) becomes

$$\|E_N\| \leq \|E_0\| + \sum_{j=0}^{N-1} (\|W_j\| + h\|B_j\|\,\|E_j\|).$$

By Corollary 1.6.2 we obtain

$$\begin{aligned}\|E_N\| &\leq \|E_0\| \exp\left(h \sum_{j=0}^{N-1} \|B_j\|\right) + \sum_{j=0}^{N-1} \|W_j\| \exp\left(h \sum_{\tau=j+1}^{N-1} \|B_\tau\|\right) \\ &\leq \|E_0\|\, e^{NLh} + \sum_{j=0}^{N-1} \|W_j\|\, e^{hL(N-j-1)},\end{aligned} \tag{6.2.5}$$

where $L = \max_{1 \leq \tau \leq N} \|B_\tau\|$. We let

$$\max_{0 \leq j \leq N} \|W_j\| = \varepsilon + \tau(h), \tag{6.2.6}$$

Then (6.2.5) becomes

$$\|E_N\| \leq \|E_0\|\, e^{LT} + \frac{e^{LT} - 1}{e^{hL} - 1} (\varepsilon + \tau(h)). \tag{6.27}$$

Usually the initial points are chosen such that

$$\lim_{N \to \infty} \|E_0\| = 0. \tag{6.2.8}$$

Hence, taking the limit as $h \to 0$, we have

$$\lim_{h \to 0} \|E_N\| \leq (e^{LT} - 1) \lim_{h \to 0} \frac{\varepsilon + \tau(h)}{e^{hL} - 1},$$

from which it is clear that if $\varepsilon \neq 0$ (the roundoff errors are considered), the right-hand side is unbounded. It is possible to derive for $\|E_N\|$ a lower bound with a similar behavior. This means that the zero solution is not practically stable.

If the roundoff errors are not considered, and if the method is consistent (see definition 2.7.1) then $\varepsilon = 0$ and $\tau(h) = O(h^{p+1})$, $p \geq 1$ and therefore

$$\lim_{h \to 0} \|E_n\| = 0 \tag{6.2.9}$$

and this implies that the null solution is totally stable.

It is possible to give a more refined analysis of (6.2.3) by considering in the limit only the components of the matrix A (see Appendix A), corresponding to the eigenvalues on the unit circle, that have nonzero limit for $n \to \infty$ (see problem 6.1).

This analysis allows us to weaken the conditions that we impose on the local error.

If, as usually happens in the modern packages of programs designed to solve the problem, one allows the stepsize to vary at each step, the difference equation for the global error is no longer autonomous even for autonomous differential equations. Gear and Tu obtain the difference equation

$$E_{n+1} = S(n)E_n + W_n, \tag{6.2.10}$$

where the matrix $S(n)$ takes into account many other factors that define the method. It is shown that

$$S(n) = \hat{S}(n) + h_n\bar{S}(n). \tag{6.2.11}$$

Suppose that the two conditions hold:

(1) $\|\Phi(n, n_0)\| \le k_0$,
(2) $\|\bar{S}(n)\| < k_1$,

where $\Phi(n, n_0)$ is the fundamental matrix of the homogeneous equation derived from (6.2.10), namely $E_{n+1} = \hat{S}(n)E_n$. The first condition (see Theorem 4.2.1) is equivalent to asking that the zero solution of

$$E_{n+1} = \hat{S}(n)E_n \tag{6.2.12}$$

is uniformly stable. The second condition is related to the boundness of the Lipshitz constant of $f(t, y)$.

From (4.6.2) we have

$$E_N = \Phi(N, n_0)E_0 + \sum_{j=n_0}^{N-1} \Phi(N, j+1)[h_j\bar{S}(j)E_j + W_j],$$

and hence it follows that

$$\|E_N\| < k_0\|E_0\| + k_0 \sum_{j=n_0}^{N-1} (h_jk_1\|E_j\| + \|W_j\|).$$

By Corollary 1.6.2 we obtain

$$\begin{aligned}\|E_N\| &\le k_0\|E_0\| \exp\left(k_0k_1 \sum_{j=n_0}^{N-1} h_j\right) + k_0 \sum_{s=n_0}^{N-1} \|W_s\| \exp\left(k_0k_1 \sum_{j=s+1}^{N-1} h_j\right) \\ &\le k_0\, e^{k_0k_1T}\left[\|E_0\| + \sum_{s=n_0}^{N-1} \|W_s\|\right],\end{aligned}$$

from which follows that if $\|E_0\|$ and $\sum_{s=n_0}^{N-1} \|W_s\|$ tend to zero as $h \to 0$, the method is convergent.

6.3. Infinite Interval

In practice one uses methods with h bounded away from zero. This means that it is not true that $h \to 0$ as n-tends to infinity. The term hB_nE_n cannot be considered as a perturbation, especially when the norm $\|B_n\|$ is very large (stiff equations). For such problems it is necessary to consider the quantity hB_nE_n as a principal component of equation (6.1.15). Because h is taken either fixed or bounded away from zero, the study of the qualitative behavior of (6.1.15) as $n \to \infty$ implies that the interval of integration is infinite. When the term hB_nE_n is taken in account, the equation is no longer linear because the elements of the matrix B_n depend on the solution. The problem becomes a difficult one. It has been studied extensively in the linear case $f(y) = \lambda y$ when $\boldsymbol{Re}\lambda < 0$. In this case, the matrix B_n becomes independent on n, namely,

$$B = \frac{\lambda}{1 - h\lambda\beta_k} \phi_k \tilde{b}^T, \tag{6.3.1}$$

where

$$\tilde{b}^T = (\beta_0 - \beta_k\alpha_0, \ldots, \beta_{k-1} - \beta_k\alpha_{k-1})^T, \tag{6.3.2}$$

and the equation (6.1.15) becomes

$$E_{n+1} = (A + hB)E_n + W_n \equiv CE_n + W_n, \tag{6.3.3}$$

whose solution is, taking $n_0 = 0$,

$$E_n = C^nE_0 + \sum_{j=n_0}^{n-1} C^{n-j+1} W_j. \tag{6.3.4}$$

The behavior of E_n will be dictated by the eigenvalues of the matrix C, which are the roots of the polynomial $\pi(z, q)$ defined in Section 2.7. The set D of values of $q = h\lambda$ for which the zero solution of the unperturbed equation

$$E_{n+1} = CE_n \tag{6.3.5}$$

is stable is said to be the absolute-stability region of the method. As in the previous case, since the roundoff errors ε_j are only bounded it is more desirable, according to Theorem 4.11.1 to require the asymptotic stability of $E_n = 0$ in (6.3.5). This implies that q must be strictly inside the absolute stability region. A very large effort has been made to study the absolute stability for different α_j and β_j, that is, for different classes of methods.

Special importance is given to methods for which the absolute stability region is unbounded and contains the complex left half plane (A-stable methods) because in this case no restrictions are to be imposed on the parameter h in order to restrict $h\lambda$ to be inside that region. We cannot go into the details of this work. We are only interested in the techniques used relative to difference equations. Recently the case of nonautonomous linear equations has also been discussed. This is very important because in the modern packages of software to solve differential equations, one allows the step-size to vary at each step. In this case, the difference equation is no longer autonomous and the considerations based on the eigenvalues of the matrices $C(n)$ are no longer enough to ensure the asymptotic stability (compare the counter example given in Section 4.4). One need to impose some conditions on the norms of matrices $\|C_n\|$ in order to get information on the fundamental matrix $\Phi(n, n_0)$ of the equation

$$E_{n+1} = C(n)E_n. \tag{6.3.6}$$

Once this has been achieved, we can use the formula (4.6.5) and obtain

$$E_n = \Phi(n, n_0)E_0 + \sum_{j=n_0}^{n-1} \Phi(n, j+1)W_j. \tag{6.3.7}$$

Suppose that the zero solution of (6.3.6) is uniformly asymptotically stable, then by Theorem 4.2.2, there exist a, $\eta > 0$ and $\eta < 1$ such that

$$\|\Phi(n, n_0)\| < a\eta^{n-n_0}. \tag{6.3.8}$$

It then follows that

$$\|E_n\| \le a\eta^{n-n_0}\|E_0\| + a\sum_{j=n_0}^{n-1} \eta^{n-j-1}\|W_j\|,$$

from which one derives at once the stability of $E_n = 0$ for the perturbed equation. We have treated so far the case of a single equation. The case of autonomous systems

$$y' = Ay, \; y(t_0) = y_0, \tag{6.3.9}$$

can be treated similarly. In fact, suppose that the matrix A can be reduced to the diagonal form by means of a similarity transformation $A = T^{-1}\Lambda T$ where $\Lambda = \operatorname{diag}(\lambda_1, \lambda_2, \ldots, \lambda)$. Then the system (6.3.9) reduces to uncoupled scalar equations and the condition becomes that each $h\lambda_i$ must be inside the absolute stability region.

6.4. Nonlinear Case

The study in the nonlinear case of equation (6.1.15) is difficult and has been made for only nonlinearities of a special kind, namely, monotone. The techniques used require essentially to bound norms of the matrix $A + hB_n$. We shall be content to present only the case of the implicit Euler method and the one-leg methods, since they are good examples of the use of the equations employed here.

Let $f: \mathbf{R} \times \mathbf{R}^s \to \mathbf{R}^s$ and suppose that for $u, v \in \mathbf{R}$,

$$(u - v)^T(f(t, u) - f(t, v)) \leq \mu \|u - v\|^2, \tag{6.4.1}$$

with $\mu \leq 0$. The difference equation (6.1.15) becomes in this case

$$l_{n+1} - l_n - h[f(t_{n+1}, z_{n+1} + l_{n+1}) - f(t_n, z_n)] = w_n. \tag{6.4.2}$$

By multiplying both sides by l_{n+1}^T we have

$$\begin{aligned} \|l_{n+1}\|^2 &\leq l_{n+1}^T \cdot l_n + h\mu \|l_{n+1}\|^2 + l_{n+1}^T \omega_n \\ &\leq \|l_n\| \, \|l_{n+1}\| + h\mu \|l_{n+1}\|^2 + \|l_{n+1}\| \, \|w_n\|, \end{aligned} \tag{6.4.3}$$

where the Cauchy–Schwarz inequality

$$|a^T b| \leq \|a\| \, \|b\| \tag{6.4.4}$$

has been used. From (6.4.3) one gets

$$\|l_{n+1}\| \leq \frac{1}{1 - h\mu} \|l_n\| + \frac{1}{1 - h\mu} \|w_n\|. \tag{6.4.5}$$

By using Corollary 1.6.1, we then obtain

$$\|l_n\| \leq \left(\frac{1}{1 - h\mu}\right)^{n - n_0} \|l_0\| + \sum_{j=n_0}^{n-1} \|w_j\| \left(\frac{1}{1 - h\mu}\right)^{n-j-1}. \tag{6.4.6}$$

Since the quantity $\eta = \dfrac{1}{1 - h\mu}$ is less than one by hypothesis, the uniform stability of the solution $l_n = 0$ follows.

The previous arguments can be generalized to the class of methods defined by

$$\rho(E)z_n - hf(\sigma(E)t_n, \sigma(E)y_n) = 0. \tag{6.4.7}$$

This class of methods was proposed by Dahlquist who gave the name of one-leg methods. There is a connection between the solutions of the difference equation (6.1.2) and the solutions of (6.4.7) (See problems 6.6,

6.7). For the sake of simplicity, we shall suppose f: $\boldsymbol{R} \times \boldsymbol{R} \to \boldsymbol{R}$. In this case the quantities in (5.4.1) are scalars. The error equation for (6.4.7) is

$$\rho(E)l_n - h[f(\sigma(E)t_n, \sigma(E)z_n + \sigma(E)l_n) - f(\sigma(E)t_n, \sigma(E)z_n)] = w_n. \tag{6.4.8}$$

Multiplying by $\sigma(E)l_n$ one obtains

$$\sigma(E)l_n \cdot \rho(E)l_n \leq h\mu|\sigma(E)l_n|^2 + w_n \cdot \sigma(E)l_n.$$

Let us switch now to the vectorial notation (See 6.1.12) and suppose that there exists a symmetric positive definite matrix G such that is the Lyapunov function

$$V_n = E_n^T G E_n$$

satisfies the inequality

$$V_{n+1} - V_n \leq 2\sigma(E)l_n \cdot \rho(E)l_n. \tag{6.4.9}$$

It then follows that

$$\begin{aligned} V_{n+1} - V_n &\leq 2\sigma(E)l_n \cdot \rho(E)l_n \\ &\leq 2h\mu|\sigma(E)l_n|^2 + 2|w_n\sigma(E)l_n| \\ &\leq 2h\mu|\sigma(E)l_n|^2 + 2|\sigma(E)l_n||w_n|. \end{aligned}$$

By considering that (here $\|\cdot\|_2$ is used.)

$$\begin{aligned} |\sigma(E)l_n| &= |(0, 0, \ldots, \beta_k)E_{n+1} + (\beta_0, \beta_1, \ldots, \beta_{k-1})E_n| \\ &\leq \left(\sum_{i=0}^{k} \beta_i^2\right)^{1/2} (\|E_{n+1}\| + \|E_n\|), \end{aligned}$$

we have, by setting $(\sum_{i=0}^{k} \beta_i^2)^{1/2} = \dfrac{\delta}{2}$;

$$V_{n+1} - V_n \leq \delta(\|E_{n+1}\| + \|E_n\|)|w_n| + \frac{h}{v}\delta^2\mu(\|E_{n+1}\| + \|E_n\|)^2. \tag{6.4.10}$$

Since $\|E_n\| \leq \|G^{-1/2}\|\,\|G^{1/2}E_n\| = \|G^{-1/2}\|\,V_n^{1/2}$, the relation becomes

$$V_{n+1} - V_n \leq \delta\|G^{-1/2}\|(V_{n+1}^{1/2} + V_n^{1/2})|w_n| + \frac{h\mu}{2}\delta^2\|G^{-1/2}\|^2(V_{n+1}^{1/2} + V_n^{1/2})^2.$$

If $\delta\|G^{-1/2}\| = a$, one obtains

$$V_{n+1}^{1/2} - V_n^{1/2} \leq a|w_n| + \frac{h\mu a^2}{2}(V_{n+1}^{1/2} + V_n^{1/2})$$

from which follows the estimate

$$V_{n+1}^{1/2} \leq \frac{a|w_n|}{1 - h\frac{\mu a^2}{2}} + \frac{1 + h\frac{\mu a^2}{2}}{1 - h\frac{\mu a^2}{2}} V_n^{1/2}.$$

Therefore,

$$V_n^{1/2} \leq \left(\frac{1 + h\frac{\mu a^2}{2}}{1 - h\frac{\mu a^2}{2}} \right)^{n-n_0} V_0 + \frac{a}{1 - h\frac{\mu a^2}{2}} \sum_{j=0}^{n-1} \left(\frac{1 + h\frac{\mu a^2}{2}}{1 - h\frac{\mu a^2}{2}} \right)^{n-j-1} |w_j|,$$

which gives a bound for $\|E_n\|_G$.

The existence of the matrix G can be established if the method is A-stable. See Dahlquist [36].

6.5 Other Techniques

The formula (4.6.16) can be used to evaluate the global error. In fact if $x(n, n_0, x_0)$ is the solution of the unperturbed problem corresponding to (6.1.2) and $y(n, n_0, x_0)$ is the solution of the perturbed problem corresponding to (6.1.3), then the difference $E_n = y(n, n_0, x_0) - x(n, n_0, x_0)$ defined by (4.6.17) gives the global error. We have already used this result in the linear case (see (6.2.3)). In applications one uses a similar formula obtained by using the method of variation of parameters of the original differential equation.

Theorem 6.5.1. *Let*

(1) *$y(t)$ be the solution of (6.1.1);*

(2) *$y_0 = z_0, z_1, \ldots, z_n$ be some approximation of $y(t)$ at the points $t_0, t_1, \ldots, t_n$;*

(3) *$p(t)$ be a sufficiently smooth function that interpolates the points z_i so that $p(t_i) = z_i$ $i = 0, 1, \ldots, n$.*

Then $E_n = y(t_n) - z_n$ satisfies the relation

$$E_n = -\int_{t_0}^{t_n} \Phi(t_n, s, p(s))[p'(s) - f(s, p(s))]\, ds, \tag{6.5.1}$$

where $\Phi(t, t_0, y_0) = \dfrac{\partial y(t, t_0, x_0)}{\partial x_0}$.

Proof. Let $s \in [t_0, t_n]$ and $y(t_n, s, p(s))$ be the solution of (6.1.1) with initial value $(s, p(s))$. It is known in the theory of differential equations that

$$\frac{\partial y(t_n, s, p(s))}{\partial s} = -\Phi(t_n, s, p(s))f(s, p(s)),$$

and

$$\frac{\partial y(t_n, s, p(s))}{\partial p} = \Phi(t_n, s, p(s)).$$

Consider the integral

$$\int_{t_0}^{t_n} \frac{d}{ds} y(t_n, s, p(s))\, ds = y(t_n, t_n, p(t_n)) - y(t_b, t_0, z_0) = -E_n.$$

This integral also is equal to

$$\int_{t_0}^{t_n} \left[\frac{\partial y(t_n, s, p(s))}{\partial s} + \frac{\partial y(t_n, s, p(s))}{\partial p} p'(s)\right] ds$$

$$= \int_{t_0}^{t_n} \Phi(t_n, s, p(s))[p'(s) - f(s, p(s))]\, ds,$$

which completes the proof. ■

The formula (6.5.1) can be used to obtain methods with higher order, to find the order of known methods and to study new methods. See [140, 171].

6.6. *The Method of Lines*

Consider the following problems

$$\frac{\partial u(x, t)}{\partial t} = \frac{\partial^2 u(x, t)}{\partial x^2}, \; u(0, t) = u(1, t) = 0, \; u(x, 0) = g(x), \tag{6.6.1}$$

and

$$\frac{\partial u(x, t)}{\partial t} = -\frac{\partial u(x, t)}{\partial x}, \; u(0, t) = 0, \; u(x, 0) = g(x), \; x > 0. \tag{6.6.2}$$

Let us discretize the interval $(0, 1)$ by taking $x_i = ih$, $i = 0, 1, \ldots, N + 1$ and consider the vectors $U(t) = (u(x_1, t), \ldots, u(x_N, t))^T$, $G = (g(x_1), \ldots, g(x_N))^T$, where $u(x_i, t)$ are approximations of the solution along the lines $x = x_i$. We approximate the operators $\frac{\partial^2}{\partial x^2}$ and $\frac{\partial}{\partial x}$ by central differences and backward difference respectively. By introducing the matrices

$$A_N = \begin{pmatrix} -2 & 1 & 0 & \dots & 0 \\ 1 & -2 & 1 & 0 & \dots 0 \\ 0 & & & & \vdots \\ \vdots & & & & 0 \\ & & & & 1 \\ 0 & \dots & 0 & 1 & -2 \end{pmatrix}_{N\times N} \tag{6.6.3}$$

and

$$B_N = \begin{pmatrix} 1 & 0 & \dots & 0 \\ -1 & 1 & & \\ 0 & & & \vdots \\ \vdots & \ddots & & \\ 0 & \dots & 0 & -1 & 1 \end{pmatrix}_{N\times N}, \tag{6.6.4}$$

the two problems can be approximated by

$$\frac{dU}{dt} = \frac{1}{\Delta x^2} A_N U, \; U(0) = G, \tag{6.6.5}$$

and

$$\frac{dU}{dt} = -\frac{1}{\Delta x} B_N U, \; U(0) = G. \tag{6.6.6}$$

We have approximated the two problems of partial differential equations with two problem of ordinary differential equations. If we discretize the time, using for example, the explicit Euler method and try the results on absolute stability, we get two different results, namely, one correct for problem (6.6.1) and one wrong for problem (6.6.2).

In fact, the eigenvalues of A_N and B_N are

$$\lambda_k = -2 + 2\cos\frac{k\pi}{(N+1)}, \; k = 1, 2, \dots, N, \tag{6.6.7}$$

$$\mu_k = 1, \; k = 1, 2, \dots, N, \tag{6.6.8}$$

and the region of absolute stability for the explicit Euler method is

$$D = \{z \in \mathbf{C} \,\|\, z + 1| \leq 1\}. \tag{6.6.9}$$

In order to have stability of the method we must have

$$-2 \leq \frac{\Delta t}{\Delta x^2} \lambda_k \leq 0 \tag{6.6.10}$$

for the first case and

$$-2 \leq -\frac{\Delta t}{\Delta x} \mu_k \leq 0 \tag{6.6.11}$$

for the second case.

From (6.6.10) and (6.6.7) it follows that the condition of stability, for the first problem is

$$\frac{\Delta t}{\Delta x^2} \leq \frac{1}{2}, \tag{6.6.12}$$

and for the second one is

$$\frac{\Delta t}{\Delta x} \leq 2. \tag{6.6.13}$$

It happens that (6.6.12) agrees with the Courant, Friedrichs and Lewy condition, while (6.6.13) does not agree with the correct condition, which is

$$\frac{\Delta t}{\Delta x} \leq 1. \tag{6.6.14}$$

The reason of the discrepancy lies in the fact that the ordinary differential equations (6.6.5) and (6.6.6) depend on $\boldsymbol{N}$, which tends to infinity when Δx tends to zero. This implies that when we study the error equation, for Δx and Δt tending to zero, the dimension of the space $\boldsymbol{R}^N$ and therefore of the matrices A_N and B_N increases. We shall see in the next section how to obtain the right conditions.

6.7. Spectrum of a Family of Matrices

Consider a family of matrices $\{C_n\}_{n \in N^+}$.

Definition 6.7.1. The spectrum of a family of matrices $\{C_n\}$ is the set of complex numbers λ such that for any $\varepsilon > 0$, there exists $n \in \boldsymbol{N}$ and $x \in \boldsymbol{C}^n$, $x \neq 0$ such that

$$\|C_n x - \lambda x\| \leq \varepsilon \|x\|. \tag{6.7.1}$$

The spectrum of $\{C_n\}$ will be denoted by $S(\{C_n\})$, or S when no confusion will arise. It is obvious that the set Σ of all eigenvalues of all matrices of the family is contained in S. We now list some properties of S, without proof.

Proposition 6.7.1. *S is a closed set.*

Proposition 6.7.2. *If the matrices are normal, then $S = \bar{\Sigma}$, where $\bar{\Sigma}$ is the closure of Σ.*

Proposition 6.7.3. *If P is a compact set of the complex plane and if $P \cap S = \phi$, then*

$$\sup_{\substack{n \in N^+ \\ z \in P}} \|(Iz - C_n)^{-1}\| < \infty \tag{6.7.2}$$

Proposition 6.7.4. *Let S be compact and $S \subset \Omega$ where Ω is an open simply connected set of C. Moreover let f be an analytic function in Ω. Then*

$$S(\{f(C_n)\}) = f(S(\{C_n\})) \tag{6.7.3}$$

Proposition 6.7.5. *Let*

$$q = \sup_{\lambda \in S} |\lambda|. \tag{6.7.4}$$

If $q < 1$, one has for every n, $m \in N^+$

$$\|C_n^m\| < k, \tag{6.7.5}$$

where k is independent on m and n.

Let us now find the spectrum of the family $\{C_n\}$:

$$C_n = \begin{pmatrix} \alpha & \gamma & 0 & \dots & & 0 \\ \beta & \alpha & \gamma & & & \vdots \\ 0 & & & \ddots & & \\ & & \ddots & & & \\ \vdots & & \ddots & & & 0 \\ & & & & & \gamma \\ 0 & \dots & & 0 & \beta & \alpha \end{pmatrix}, \tag{6.7.6}$$

where α, $\dot{\times}$, γ are real and we shall suppose for simplicity that $|\beta/\gamma| < 1$. Let be $Q = S \backslash \Sigma$. The elements of Q will be called quasi-eigenvalues.

Theorem 6.7.2. *The set of complex numbers λ such that the zero solution of the difference equation*

$$\begin{gathered} \gamma x_{i+1} + (\alpha - \lambda) x_i + \beta x_{i-1} = 0 \\ x_1 = 1, \qquad x_2 = \frac{\lambda - \alpha}{\gamma} \end{gathered} \tag{6.7.7}$$

is asymptotically stable, is contained in $S(\{C_n\})$.

Proof. The characteristic polynomial of (6.7.7) is

$$\gamma r^2 + (\alpha - \lambda)r + \beta = 0 \tag{6.7.8}$$

and to get asymptotic stability it must be a Schur polynomial (see Theorem 2.6.1). Suppose it is, then taking as vector x the vector made up with the first n components of the solution of problem (6.7.7), one has $C_n x - \lambda x = (0, 0, \ldots, \gamma x_{n+1})^T$. It follows that $\|C_n x - \lambda x\| = |\gamma||x_{n+1}| = |\gamma||c_1 r_1^{n+1} + c_2 r_2^{n+1}|$, where r_1 and r_2 are the roots of (6.7.8) which, for simplicity, are supposed simple and c_1, c_2 are deduced from the initial conditions, that is

$$c_1 = \frac{1}{r_1 - r_2}, \qquad c_2 = \frac{1}{r_2 - r_1}.$$

We then get

$$\|C_n x - \lambda x\| \leq \frac{1}{|r_1 - r_2|} |r_1^{n+1} + r_2^{n+1}| \, \|x\|,$$

where we have considered that $\|x\| \geq 1$. The quantity multiplying $\|x\|$ can be made arbitrary small if $|r_1|$ and $|r_2|$ are less than 1. Similar considerations can be made if $|r_1| = |r_2|$. One can show that if $|r_1|$ or $|r_2|$ are greater than 1, the corresponding λ is no part of the spectrum and $\lambda \notin \Sigma$ (see Problem 6.9). ■

The next theorem explicitly furnishes the boundary of S.

Theorem 6.7.2. *The boundary of S is given by*

$$x = \alpha + (\beta + \gamma) \cos \theta, \qquad y = (\gamma - \beta) \sin \theta, \tag{6.7.9}$$

where $\lambda = x + iy$, and $0 \leq \theta \leq \pi$.

Proof. The proof reduces to find for which values of λ the polynomial (6.7.8) is a Schur polynomial. This can be done easily by using Theorem B1 of Appendix B, obtaining the conditions (6.7.9). ■

The set S has then as boundary, the ellipses with center $(\alpha, 0)$ and semiaxes $(\beta + \gamma)$ and $(\gamma - \beta)$.

By considering that the eigenvalues $\lambda_k^{(n)}$ of the matrices C_n are given by $\lambda_k^{(n)} = \alpha + 2(\beta\gamma)^{1/2} \cos \dfrac{k\pi}{n+1}$, it follows that $\lambda_k^{(n)} \in S$.

Particular cases are:

(a) $\alpha = -2$, $\beta = \gamma = 1$; in this case the family $\{C_n\}$ coincides with the family $\{A_n\}$ defined by (6.6.3). The spectrum $S(\{A_n\})$ is then the segment

$$-2 + 2\cos\theta, \qquad 0 \leq \theta \leq \pi. \tag{6.7.10}$$

(b) $\alpha = 1, \beta = -1, \gamma = 0$; in this case the family $\{C_n\}$ coincides with the family $\{B_n\}$ defined by (6.5.4). The spectrum $S(\{B_n\})$ is then the circle defined by

$$x = 1 - \cos\theta, \ y = -\sin\theta. \tag{6.7.11}$$

Once the spectrum of the family has been obtained, the correct stability conditions can be obtained. If the case where the ordinary differential equations are discretized by using the Euler method for example, instead of (6.6.10) and (6.6.11), one imposes

$$\frac{\Delta t}{\Delta x^2} S(\{A_n\}) \subset D \tag{6.7.12}$$

for the first problem and

$$\frac{\Delta t}{\Delta x} S(\{B_n\}) \subset D \tag{6.7.13}$$

for the second problem. In fact we have:

Theorem 6.7.3. *Suppose that the problems* (6.6.5) *and* (6.6.6) *are solved with the explicit Euler method. If* (6.7.12) *and* (6.7.13) *hold, then the resulting difference equations are stable.*

Proof. Let V_s, $s = 0, 1, \ldots$ be an approximation of $U(t_s)$, the Euler method will produce the difference equations

$$U_{s+1} = U_s + \frac{\Delta t}{\Delta x^2} A_n U_s = \left(I + \frac{\Delta t}{\Delta x^2} A_n\right) U_s$$

and

$$U_{s+1} = \left(I - \frac{\Delta t}{\Delta x} B_n\right) U_s$$

respectively. When (6.7.12) and (6.7.13) hold, by proposition 6.7.3, it follows that the spectrum of the families of matrices $\left\{I + \frac{\Delta t}{\Delta x^2} A_n\right\}$ and $\left\{I - \frac{\Delta t}{\Delta x} B_n\right\}$ are contained in the unit circle. This implies, by Proposition 6.7.4, that all the powers of the matrices of the families are uniformly bounded and this implies the stability. One should consider, however, that the condition of consistency for the space discretization implies $\alpha + \beta + \gamma = 0$ and the origin is a point of the boundary of both S and ∂D. This implies that (6.7.12) and (6.7.13) cannot be strictly satisfied. For some class of matrices, for example normal matrices, this does not create any problem. Note that condition (6.7.12) is not more restrictive than the one obtained by using only the eigenvalues because in this case, the matrices are normal and the spectrum is only the closure of the set of all the eigenvalues. Condition (6.7.13) is more restrictive because the spectrum is a disk in the complex plane, while the set of eigenvalues is made up of only one point. More generally to each method will be associated a region of absolute stability D defined by the discretization of t and a region S, which is a function of the spectrum of a family of matrices defined by the discretization of the space variables. For the stability the second region must be contained in the first. This approach is also useful because it permits the choice of appropriate discretization of the time as consequence of the discretization of the space variables. ■

6.8. Problems

6.1. Suppose that the matrix A has the eigenvalues $1 = \lambda_1 > \lambda_2 \geq \ldots \geq \lambda_k$. Using the decomposition (A2.2) of the matrix, give a bound for $\|E_n\|$ in (6.2.3).

6.2. Show that for the Euler method $y_{n+1} = y_n + hf_n$ the region of absolute stability is the circle with center at -1 and radius 1.

6.3. Show that the region of absolute stability of the trapezoidal method $z_{n+1} = z_n + \frac{h}{2}(f_n + f_{n+1})$ is the left-hand plane.

6.4. Find the one-leg correspondent of the trapezoidal method. It is known as the implicit midpoint method.

6.5. Show that the region of absolute stability of the midpoint method $z_{n+2} = z_n + 2hf_{n+1}$ is the segment $[-i, i]$ of the imaginary axis.

6.6. Suppose that y_n satisfies the one-leg equation $\rho(E)y_n - hf(\sigma(E)y_n) = 0$. Show that $\tilde{y}_n = \sigma(E)y_n$ satisfies the equation $\rho(E)\tilde{y}_n - h\sigma(E)f(\tilde{y}_n) = 0$. (Hint: the operators ρ and σ commute.)

6.7. Suppose that $\{y_n\}$ satisfies the equation $\rho(E)y_n - h\sigma(E)f(y_n) = 0$ and $P(z)$ and $Q(z)$ are polynomials satisfying $P(z)\sigma(z) - Q(z)\rho(z) = 1$. Then show that $\tilde{y}_n = P(E)y_n - hQ(E)f(y_n)$ satisfies the one-leg equation $\rho(E)\tilde{y}_n - hf(\sigma(E)\tilde{y}_n) = 0$.

6.8. Suppose that $c_n < -\mu$ $(\mu > 0)$ for all n, $\beta_n \neq 0$ in (6.1.10) and the roots of $\sigma(z)$ are inside the limit disk. Show that the error e_n tends to zero.

6.9. Show that in Theorem 6.7.1 if one of the roots of (6.7.7) has modulus greater than 1, then λ does not lie in Q.

6.10. Show that for the matrix (6.7.6) it is $\|C_n x - \lambda x\| = |\beta\rho^{n-1} U_n(q)|$, where $\rho = (\beta/\gamma)^{1/2}$ and $q = (\lambda - \alpha)/2(\beta\gamma)^{1/2}$ and $U_n(q)$ is the Chebyshev polynomial of second species. (Hint: use the results of Section 2.3.)

6.11. Show that the midpoint method cannot be used for the time discretization of 6.6.5.

6.12. Consider the second order hyperbolic equation $U_{tt} - U_{xx} = 0$, $U(x, 0) = f(x)$, $U(0, t) = u(\pi, t) = 0$ and discretize the space variable obtaining a system of first order equations, $\dfrac{dw}{dt} = \dfrac{1}{\Delta x^2} D_{2N} w$, $w(0) = \psi$, where $w = (U, V)^T$, $\psi = (\Phi, 0)$, $U, V, \Phi \in \boldsymbol{R}^N$, N is the number of discrete points in $(0, \pi)$, and

$$D_{2N} = \begin{pmatrix} 0 & \Delta x^2 I \\ A_N & 0 \end{pmatrix}.$$

6.13. Show that D_{2N}^2 is normal and find its spectrum. Using this result, show that the midpoint method can be used in this case.

6.9. Notes

The problem of numerical methods for differential equations has been extensively studied in the last 30 years. For important works in this field see Henrici [74, 75] and Dahlquist [33]. The study of nonlinear case was also initiated systematically by Dahlquist [39, 34, 35, 36, 37, 38]. Important contributions have also been made by Odeh and Liniger [11], Nevanlinna and Liniger [118], Nevanlinna and Odeh [117]. For the class of Runge–Kutta methods, not treated in this book, see Butcher [20], Burrage and Butcher [17], [18]. For applications of the theory of difference equations to methods for boundary value problems see Mattheij [105]. Material on the spectrum of family of matrices can be found in Bakhvalov [10], Godunov and Ryabencki [59], Di Lena and Trigiante [45]. A general treatment in Hilbert space setting of the spectrum of infinite Toeplitz matrices can be found in Hartman and Wintner [72], as well as in the Toeplitz original papers [165]. Applications of the enlarged spectrum of symmetric tridiagonal matrices (Jacobi matrices) are usual in the theory of orthogonal polynomials, see for example Matè and and Nevai [101] and the references therein.

CHAPTER 7 Models of Real World Phenomena

7.0. Introduction

In this chapter we offer several examples of real world models to illustrate the theory developed, to show how versatile difference equations are and to demonstrate how often the results of discrete models can explain observed phenomena better when compared to continuous models. Furthermore, the contents of this chapter are also intended to help the practitioners who can look first at the examples and the proceed to read necessary theory of difference equations.

Section 7.1 deals with linear models of population dynamics, sketches the development and discusses the appearance of population waves. We devote Section 7.2 to the study of nonlinear models of population dynamics including bifurcation and chaotic behavior. In Section 7.3, we consider models arising in the distillation of a binary ideal mixture of two liquids while in Section 7.4, we discuss models from theory of economics. Models from queueing theory and traffic flow are investigated in Section 7.5 and in Section 7.6 we collect some interesting examples dealing with various models.

7.1. Linear Models for Population Dynamics

The dynamics of populations have been extensively studied since Malthus. The object of study is the evolution in time of a group of individuals who follow the simplest and essential life events: they are born, mature, reproduce and die. The group can be of bacteria or of animals or of humans. For simplicity, in all the models one usually considers only the female part of the population. Both continuous and discrete models have been proposed. More than the law of mechanics, which could have been written in discrete form, it seems, however, not always justified to use in this field continuous models. In fact the usual assumption that the variation per unit time of the number of individuals is small compared to the total number is no longer valid as in mechanics. Often the results of continuous and discrete models are qualitatively similar, but sometimes, especially in nonlinear models they are not and the continuous models are unable to explain observed phenomena (see Section 7.2).

Let y_n be the number of females of a population at time t_n. The simplest model, called after Malthus, Malthusian model, is the following

$$y_{n+1} = ay_n, \qquad y_{n_0} = y_0, \qquad a > 0, \tag{7.1.1}$$

which simply states that the number of newborn individuals as well as the number of individuals that are deceased in the time interval Δt_n is proportional to the number of individuals at time t_n. The parameter a is called intrinsic growth rate and is simply the difference between the birth and death rates.

The solution of (7.1.1) is $y_n = a^{n-n_0}y_0$ and has been found to be very effective to fit exponential data for not large populations. One of the weaknesses of the model is that for $n \to \infty$, $y_n \to \infty$, which is impossible due to limitation of the resources.

The foregoing simple model does not take into account the age structure of the population, which is important because all of the essential life events depend on the age of individuals. A more refined model is the following (often called Leslie model). Let us divide the lifetime L into M parts and then population in groups (or classes) $y_1(n)$, $y_2(n), \ldots, y_M(n)$ such that the group $y_i(n)$ is made of individuals whose age a_i satisfies $(i-1)\dfrac{L}{M} \le a_i < i\dfrac{L}{M}$.

Considering the unit time equal to the time spanned by a class, one has that $y_{i+1}(n+1)$ is made up of all the individuals in the class i at time n except of those who are deceased (or removed in some way). That is,

$$y_{i+1}(n+1) = \beta_i y_i(n), \qquad i = 1, 2, \ldots, M-1,$$

where $\beta_i > 0$ is the survival rate of the ith age and

$$y_1(n+1) = \sum_{i=1}^{M} \alpha_i y_i(n),$$

which states the newborns at time $n+1$ as sum of the contribution of the different groups each of which is considered to have an homogeneous behavior in the reproduction. Having the human population in mind, one would expect α_i to be zero in the extreme groups and the maximum in some group in the middle. In nature, however, there is a wide range of possibilities; for example $\alpha_i = 0$ for all groups except one, giving rise to different kinds of behavior.

By introducing the vector $y(n) = (y_1(n), \dots, y_M(n))^T$ and the matrix

$$A = \begin{pmatrix} \alpha_1 & \beta_2 & & & \alpha_M \\ \beta_1 & 0 & & & 0 \\ 0 & & & & \\ \vdots & & & & \\ 0 & & 0 & \beta_{M-1} & 0 \end{pmatrix}_{M\times M}, \tag{7.1.2}$$

the model can be written as

$$y(n+1) = Ay(n), \qquad y(n_0) = y_0, \tag{7.1.3}$$

where y_0 is supposed to be known. The solution is $y(n) = A^n y_0$ (we take $n_0 = 0$ for simplicity). The behavior of the solution will then follow from the spectral properties of the matrix A.

Consider first the case $\alpha_i > 0$ and $\beta_i > 0$ for all indices. In this case it is easy to verify that $A^M > 0$ (see sec. A.6). By Perron–Frobenius theorem (see Theorem A.6-2), it follows that A has a real simple eigenvalue λ_1 associated with a positive eigenvector u_1. The remaining eigenvalues are inside the circle B_{λ_1} in the complex plane. By using the expression (A2.1-6) for the powers of A and considering that in the present case $m_1 = 1$, one has

$$A^n = \lambda_1^n\left[Z_{11} + \sum_{k=2}^{s}\sum_{i=0}^{m_k-1}\frac{\lambda_k^{n-i}}{\lambda_1^n}\binom{n}{i}Z_{ki}\right],$$

from which it follows that the double sum in brackets tends to zero as $n \to \infty$. Considering that Z_{11} is the projection to eigenspace corresponding to λ_1, one has $A^n y_0 \to c\lambda^n u_1$, where c is a constant.

The population tends to grow as λ_1^n and is close to u_1, that means that the distribution between the age groups tends to remain fixed. In fact,

$$\frac{y_i(n)}{\sum_{i=1}^{M} y_i(n)} \to \frac{u_i}{\sum u_i}. \tag{7.1.4}$$

The vector u_1 is called the stable age vector and λ_1 the natural growth rate. Of course the overall population will grow for $\lambda_1 > 1$ and will extinguish, for $\lambda_1 < 1$.

Special cases of interest are those where some α_i can be zero. To make the study simpler, it is useful to change variables. Letting $l_1 = l$, $l_k = \prod_{i=0}^{k-1} \beta_i$ $(k = 2, \ldots, M)$, $a_i = l_i\alpha_i$ $(i = 1, 2, \ldots, M)$, $D = \text{diag}(l_1, l_2, \ldots, l_M)$, $x(n) = D^{-1}y(n)$ and $B = D^{-1}AD$, one verifies that

$$B = \begin{pmatrix} a_1 & & a_2 & \ldots & a_m \\ 1 & 0 & \ldots & 0 & \\ 0 & \ddots & \ddots & & \vdots \\ \vdots & & & & \\ 0 & \ldots & 0 & 1 & 0 \end{pmatrix},$$

which is in the companion matrix (see section 3.3).

The model is now

$$x(n+1) = Bx(n), \tag{7.1.5}$$

and the characteristic equation of B is then (see 3.3.3):

$$\lambda^M - a_1\lambda^{M-1} - a_i\lambda^{M-2} \ldots a_{M-1}\lambda - a_M = 0. \tag{7.1.6}$$

The following theorem, due to Cauchy, is very useful in this case.

Theorem 7.1.1. *If a_i is nonnegative and the indices of the positive a_i have the greatest common divisor* 1, *then there is a unique simple positive root λ_1 of the greatest modulus.*

For a proof, see Ostrowski [131]. Since the eigenvector u_1 of B is $u_1 = (\lambda_1^{M-1}, \lambda_1^M, \ldots, 1)^T$, it follows that u_1 is a positive vector and all the previous results are still valid.

When the hypothesis on the indices of Theorem 7.1.1 are not satisfied, then some eigenvalues may have the same modulus of λ_1 giving rise to the interesting phenomenon of population waves. In this case, there are oscillating solutions to the model. To see this in a simpler way, let us consider the scalar equation for the population of the first group. This can be seen from (7.1.5). The first equation of the system gives

$$x_1(n+1) = \sum_{i=1}^{M} a_i x_i(n) \tag{7.1.7}$$

and from the following equations, choosing appropriately the index n, we obtain $x_i(n) = x_1(n - i + 1)$. By substitutions one has

$$x_1(n+1) = \sum_{i=1}^{M} a_i x_1(n - i + 1), \tag{7.1.8}$$

which is a scalar difference equation of order M. The characteristic polynomial coincides with (7.1.6). The solution of (7.1.8) in terms of the roots of (7.1.6) is given by (2.3.7) with λ_i instead of z_i. If there are two distinct roots with the same modulus, say $\lambda_1 = \rho\, e^{in\theta}$, $\lambda_2 = \rho\, e^{-in\theta}$, one has

$$x_1(n) = c_1\rho^n\, e^{in\theta} + c_2\rho\, e^{-in\theta} + \text{other terms}.$$

It is possible to find two new constants a and ψ such that $c_1 = a\, e^{i\psi}$, $c_2 = a\, e^{-i\psi}$ and then

$$x_1(n) = 2^n\rho a \cos(n\theta + \psi) + \text{other terms},$$

from which follows that $x_1(n)$ is an oscillating function. In the general case more than one period can coexist. In the extreme case $a_i = 0$, $i = 1, 2, \ldots, M-1$, $a_M \neq 0$ a number proportional to M of periods can coexist and for large M a phenomena similar to chaos (see next section) may appear. The population waves have been observed in insect populations.

7.2. The Logistic Equation

The discrete nonlinear model, which we are going to present for the dynamics of populations and which takes into account the limitation of resources, has been used successfully in many areas such as meteorology, fluid dynamics, biology and so on. Its interest consists in the fact that in spite of its simplicity, it presents a very rich behavior of the solutions that permits to explain and predict a lot of experimental phenomena. Let $N(n)$ be the number of individuals at time n and suppose that $N(n+1)$ is a function of $N(n)$, that is,

$$N(n+1) = F(N(n)). \tag{7.2.1}$$

This hypothesis is acceptable if two generations do not overlap. For small N, (7.2.1) must recover the Malthusian model, that is,

$$F(N(n)) = aN(n) + \text{nonlinear terms}.$$

The nonlinear terms must take into account the fact that when the resources are bounded there must be a competition among the individuals, which is proportional to the number of encounters among them. The number of encounters is proportional to $N^2(n)$. The model becomes

$$N(n+1) = aN(n) - bN^2(n), \tag{7.2.2}$$

with a and b positive. The parameter a represents the growth rate and b is a parameter depending on the environment (resources). The quantity a/b is said to be the carrying capacity of the environment. For $a > 1$, this equation has a critical point $\bar{N} = \dfrac{a-1}{b}$ which, as shown below, is globally asymptotically stable for $y_0 \in (0, 1)$ and a in a suitable interval. This means that $\lim_{n\to\infty} N(n) = \bar{N}$ and $\bar{N}$ is a limit size of the population. $\bar{N}$ depends on a and b, which can vary with the time: b can diminish because the population learns to better use the existing resources or to discover new ones; a can grow because the population learns how to live longer or how to accelerate the birth ratio. Anyway, taking into account the variation of $\bar{N}$ (which happens with a longer time scale), one sees the evolution of the population can be described as a sequence of equilibrium points with increasing values of $\bar{N}$. In case of two similar species, it can be shown that the species with longer $\bar{N}$ will survive. This is the Darwinian law of the survival of the "fittest." Almost all the previous results are similar to those obtained in the continuous case. A further analysis of the logistic equation shows solutions that are unexpected if one thinks the discrete equation merely as an approximation of the continuous one. To see this, let us change the variable and simplify the equation. Let $y_n = \dfrac{a}{b} N(n)$. The model becomes

$$y_{n+1} = a y_n (1 - y_n) \equiv f(y_n). \tag{7.2.3}$$

The new variable y_n represents the population size expressed in unit of carrying capacity a/b.

The equation (7.2.3) has two critical points (the term equilibrium points is used more in this context), the origin and $\bar{y} = (a-1)/a$, which has physical meaning only for $a > 1$. For $a < 1$, by using the theorems of stability by first approximation (see Section 4.7), one sees at once that the origin is asymptotically stable. It can be shown (see problem 7.3), that it is asymptotically stable for λ_0 in the interval $(0, 1)$. For $a > 1$, the second critical point becomes positive and the origin becomes unstable. The stability of $\bar{y}$ can be studied locally by the linearization methods using again one of the theorems of Section 4.7. In fact $y_{n+1} - \bar{y} = f'(\bar{y})(y_n - \bar{y}) + \dfrac{f''(\bar{y})}{2}(y_n - \bar{y})^2 = (2-a)(y_n - \bar{y}) - a(y_n - \bar{y})^2$.

If the coefficient of the linear term is less than one in absolute value then $\bar{y}$ is asymptotically stable.

As a consequence, one has that for $1 < a < 3$ the point $\bar{y}$ is asymptotically stable. For $a = 2$, the solution can be found explicitly (see Problem 1.2.7).

For $a > 3$, a new phenomenon appears. There exists a couple of points x_1 and x_2 such that

$$x_2 = f(x_1) \quad \text{and} \quad x_1 = f(x_2). \tag{7.2.4}$$

The couple x_1 and x_2 is called a cycle of period two. From (7.2.4) it follows that

$$x_1 = f(f(x_1)) = f^{(2)}(x_1), \tag{7.2.5}$$

that is, x_1 and x_2 are fixed points of the difference equation

$$y_{n+2} = f^{(2)}(y_n). \tag{7.2.6}$$

One can determine the fixed points by using the Equation (7.2.5), which is a fourth degree equation (it contains as roots the origin and $\bar{y}$, which are circles of any period). The following theorem permits us to simplify the problem. Let $f[x, y]$ be the usual divided difference, that is,

$$f[x, y] = \frac{f(x) - f(y)}{x - y}.$$

Theorem 7.2.1. *The solutions of period two for the difference equation $y_{n+1} = f(y_n)$ exist iff there exist values of $x \in (0, 1)$ such that $f[s, f(x)] = -1$.*

Proof. By definition

$$f[x, f(x)] = \frac{f(x) - f^{(2)}(x)}{x - f(x)}. \tag{7.2.7}$$

If $f^{(2)}(x) = x$ then (7.2.7) is equal to -1. On the other hand, if (7.2.7) is equal to -1, then it follows at once that $x = f^{(2)}(x)$. ■

Applying the previous results one obtains in the present case $f[x, f(x)] \equiv a(1 - x - ax + ax^2) = -1$.

This equation has real roots for $a \geq 3$. It can be shown, for example, considering the linear part of $f^{(2)}(x)$, that the circle is asymptotically stable for values of a in a certain range, while $\bar{y}$ becomes unstable. One says in this case that the solution $\bar{y}$ bifurcates in the circle of order two. As a result, it follows that if $\bar{y}$ is greater than $\frac{2}{3}$ (in units of carrying capacity) then the system tends to oscillate between two values. We cannot go into details on what happens for $a > 3$ (see [106] for details), but we shall qualitatively sketch the picture. For higher values of a, a cycle of order four appears and then a cycle of order eight and so on all the cycles of even period 2^n will appear. All this happens as a goes from 3 to 3.57, where the last point is an accumulation point of cycles of 2^n periods. What is the solution like to the left of this point?

For a near 3.57 one can find a very large number of points that are parts of some solution of even period (which can be very large and hardly distinguishable from the periodic solutions). In this region even if two similar populations evolve starting from two very close initial points, their history will be completely different. After a long time every subinterval of $(0, 1)$ will contain a point of the trajectory and if one maps the density of the number of occurrences of y_n in subintervals of $(0, 1)$ the pictures are very similar to sample functions of stochastic processes. This behavior has been called chaos. After the value 3.57 of a, the solution of period three appears. There is a result of Li and Yorke that states if there is a cycle of period three, then there are solutions of any integer periods and for the same reasons discussed above, the term chaos is appropriate in this case as well.

Almost all of the previous results can be extended to more general functions f leading to the conclusion that the qualitative results are widely independent from the particular function chosen to describe the discrete model.

7.3. Distillation of a Binary Liquid

The distillation of a binary ideal mixture of two liquids is often realized by a set of N-plates (column of plates) at the top of which there is a condenser and at the bottom there is a heater. At the base of the column there is a feeder of new liquid to still. A stream of vapor, whose composition becomes richer from the more volatile component, proceeds from one plate to the next one until it reaches the condenser from which part of the liquid is removed and part returns to the last plate. On each plate, which is at different temperature, the vapor phase will be in equilibrium with a liquid phase. Because of this, a liquid stream will proceed from the top to the bottom. We suppose that the liquids are ideal (that is the Raoult's law applies) as well as the vapors (that is the Dalton's law applies).

Let y_i $(i = 1, 2)$ be the mole fraction of the ith component in the vapor phase and x_i the mole fraction of the same component in the liquid phase. Of course $y_1 + y_2 = 1$ and $x_1 + x_2 = 1$ (the sum of the mole fraction in each phase is 1 by definition).

In this hypothesis the quantity, called relative volatility,

$$\alpha = \frac{y_1}{y_2}\frac{x_2}{x_1}$$

can be considered constant in a moderate range of the temperature (see [173]). If $\alpha > 1$ one says that the first component is more volatile. For simplicity, we shall consider as reference only the more volatile component. Setting $y_1 = y$ and $x_1 = x$, one has

$$\alpha = \frac{y(1-x)}{x(1-y)}, \tag{7.3.1}$$

which will be considered valid every time the two phases are in equilibrium.

Let us see what happens on the nth plate. Here the two components are in equilibrium in the two phases. Let x_n be the mole fraction of the more volatile component in the liquid phase, y_n^* the mole fraction in vapor phase of the same component, and y_n the mole fraction of the same component leaving the plate n. If we assume that the plate efficiency is 100 percent, then $y_n^* = y_n$. Moreover, part of the liquid will fall down with mole rate d and the vapor will go up with mole rate V. Let D be the mole rate of the product, which is withdrawn from the condenser. Consider now the system starting from the nth plate (above the point where new liquid enters into the apparatus) and the condenser. We can write the balance equation

$$Vy_{n-1} = dx_n + Dx_D, \tag{7.3.2}$$

where x_D is the mole fraction of the liquid withdrawn from the condenser. To this equation one must add the definition of relative volatility that will hold for the equilibrium of the two phases at each plate

$$\alpha = \frac{y_n(1-x_n)}{x_n(1-y_n)}. \tag{7.3.3}$$

From the last relation we obtain

$$y_n = \frac{\alpha x_n}{1+(\alpha-1)x_n}$$

and, after substitution in (7.3.2) we get

$$x_n x_{n-1} + \frac{x_n}{\alpha-1} + \frac{Dx_D(\alpha-1) - V\alpha}{d(\alpha-1)} x_{n+1} + \frac{Dx_D}{d(\alpha-1)} = 0, \tag{7.3.4}$$

from which, by letting

$$a = \frac{1}{\alpha-1}; \qquad b = \frac{Dx_D(\alpha-1) - V\alpha}{d(\alpha-1)}; \qquad c = \frac{Dx_D}{d(\alpha-1)}$$

one obtains

$$x_n x_{n-1} + ax_n + bx_{n-1} + c = 0. \tag{7.3.5}$$

This is a difference equation of Riccati type (see 1.5.8).

Let us consider the boundary conditions. The initial condition depends on how the apparatus is fed from the bottom and we will leave this undetermined. The other boundary condition will depend on the condition that we impose on the composition of the fluid on some plate. In the literature there are two different types of such conditions. The first one (see [19]) requires that there is a plate i for which $x_{i-1} = x_i$. We shall show that this condition either leads to a constant solution or to no solutions (according to the initial conditions). The second boundary condition (which can be deduced from Fig. 16 of [20]) requires that $y_N = x_D$, $x_{N+1} = x_D$.

Let us start with the first condition, that is $x_i = x_{i-1}$ for some $i \in [2, N]$. Then x_i must satisfy the equation (obtained from (7.3.5))

$$x_i^2 + (a + b)x_i + c = 0. \tag{7.3.6}$$

Now the Riccati equation can be transformed to a linear one by setting

$$x_n = \frac{z_n}{z_{n-1}} - a \tag{7.3.7}$$

to obtain

$$z_{n+1} + (b + a)z_n + (c - ab)z_{n-1} = 0. \tag{7.3.8}$$

If λ_1 and λ_2 are two solutions of the polynomial associated with the previous equation one has (if $\lambda_1 \neq \lambda_2$)

$$z_n = c_1\lambda_1^n + c_2\lambda_2^n$$

and therefore

$$x_n = \frac{c_1\lambda_1^n + c_2\lambda_2^n}{c_1\lambda_1^{n-1} + c_2\lambda_2^{n-1}} - a. \tag{7.3.9}$$

Put $n = i$ and $n = i - 1$. The only way to obtain $x_i = x_{i-1}$ is taking either $c_1 = 0$ or $c_2 = 0$, that is, $x_n = \lambda_1 - a$ or $x_n = \lambda_2 - a$. One verifies at once that these two solutions satisfy (7.3.6).

If the initial condition does not match one of the two constant solutions, then there does not exist any solution.

Concerning the second boundary conditions, which is consistent with the condition $V = d + D_i$ necessary for other physical considerations, one has

$$Vy_N = Lx_{N+1} + Dx_D.$$

Imposing $y_N = x_D$ we get

$$x_D = \frac{c_1\lambda_1^{N+1} + c_2\lambda_2^{N+1}}{c_1\lambda_1^N + c_2\lambda_2^N} - a \equiv \frac{\dfrac{c_1}{c_2}\lambda_1^{N+1} + \lambda_2^{N+1}}{\dfrac{c_1}{c_2}\lambda_1^N + \lambda_2^N} - a. \tag{7.3.10}$$

One obtains an equation whose unknowns are $\frac{c_1}{c_2}$ (the Riccati equation is of first order and its general solution must contain only an arbitrary constant) and N (the number of plates needed to complete the process).

Let us put, for simplicity, $K = \frac{c_1}{c_2}$. One obtains, after some simple algebraic manipulations,

$$N = \frac{\log \frac{1}{k}\left(-\frac{\lambda_2 - a - x_D}{\lambda_1 - a - x_D}\right)}{\log \frac{\lambda_1}{\lambda_2}}. \tag{7.3.11}$$

One can verify that in order to obtain $x_D = 1$ (at this value only the more volatile component is present in the distillate), one needs $N = \infty$. The value of K is determined from the condition of the feed plate (which can be assumed as the zeroth plate).

7.4. Models from Economics

A model called the cobweb model concerns the interaction of supply and demand for a single good. It assumes that the demand is a linear function of the price at the same time, while the supply depends linearly on the price at the previous period of time. The last assumption is based on the fact that the production is not instantaneous, but requires a fixed period of time. Let the above period of time be unitary. We shall then have

$$d_n = -ap_n + d_0,$$
$$s_n = bp_{n-1} + s_0,$$

where d_n, s_n are respectively the demand and supply functions and a, b, d_0, s_0 are positive constants.

This model is based on the following assumptions:

(1) The producer hopes that the price will be the same in the next period and produces according to this.
(2) The market determines at each time the price such that

$$d_n = s_n. \tag{7.4.2}$$

From (7.4.1) and (7.4.2) one obtains

$$p_n = \left(-\frac{b}{a}\right)p_{n-1} + \frac{d_0 - s_0}{a}.$$

The equilibrium price p_e is obtained by setting $p_n = p_{n-1} = p_e$, which gives

$$p_e = \frac{d_0 - s_0}{a + b}$$

and the solution of (7.4.3) is given by

$$p_n = \left(-\frac{b}{a}\right)^n p_0 + \left[1 - \left(-\frac{b}{a}\right)^n\right] p_e, \tag{7.4.4}$$

which is deduced from (1.5.4).

If $\frac{b}{a} < 1$, it follows that the equilibrium price p_e is asymptotically stable, that is $\lim_{n\to\infty} p_n = p_e$. Since $-\frac{a}{b}$ is negative, we see that p_n will oscillate approaching p_e. For $\frac{a}{b} > 1$ the equilibrium price is unstable and the process will never reach this value (unless $p_0 = p_e$).

As a more realistic model, take s_n as a linear function of a predicted price at time n. This means that the producer tries to predict the new price using information on the past evolution of the price. For example, he can assume as the new price $p_n = p_{n-1} + \rho(p_{n-1} - p_{n-2})$ obtaining

$$s_n = b(p_{n-1} + \rho(p_{n-1} - p_{n-2})) + s_0,$$

while the first equation remains unchanged.

By equating the demand and supply, one obtains

$$-ap_n + d_0 = +bp_{n-1} + b\rho b_{n-1} - b\rho p_{n-2} + s_0$$

from which

$$p_n + \frac{b}{a}(1 + \rho)p_{n-1} - \frac{b}{\alpha}\rho p_{n-2} + \frac{s_0 - d_0}{a} = 0. \tag{7.4.6}$$

The equilibrium price is now $p_e = \frac{d_0 - s_0}{a - b}$.

The homogeneous equation is

$$p_n + \frac{b}{a}(1 + \rho)p_{n-1} - \frac{b}{a}\rho p_{n-2} = 0 \tag{7.4.7}$$

and the characteristic polynomial reduces to

$$z^2 + \frac{b}{a}(1 + \rho)z - \frac{b\rho}{a} = 0. \tag{7.4.8}$$

Let z_1 and z_2 be the roots of this equation. The general solution of the homogeneous equation is

$$p_n = c_1 z_1^n + c_2 z_2^n,$$

while the general solution of (7.4.6) is

$$p_n = c_1 z_1^n + c_2 z_2^n + p_e, \tag{7.4.9}$$

where c_1 and c_2 are determined by using the initial conditions. From (7.4.9) it follows that if both $|z_1|$ and $|z_2|$ are less than one, one has $\lim_{n\to\infty} p_n = p_e$. If at least one of the two roots has modulus greater than one, then $\lim_{n\to\infty} p_n = \infty$ (for generic initial conditions). The problem is now reduced to derive the conditions on the coefficients of the equation in order to have $|z_1|$ and $|z_2|$ both less than one. A necessary condition is $\left|\frac{b\rho}{a}\right| < 1$, since $\frac{b\rho}{a} = -z_1 z_2$. The other condition is (see Theorem B1)

$$\left|\frac{\frac{b}{a} + \frac{b\rho}{a}}{1 - \frac{b\rho}{a}}\right| < 1,$$

which is satisfied for $\frac{a}{b} < 3$ and $-\frac{a}{b} < \rho < \frac{a-b}{2b}$.

The positive value of ρ means that the producer expects the price will continue to have the same tendency, while the negative value means that the producer expects to have an inversion in the tendency.

To an equation similar to (7.4.6), one arrives in the Samuelson's model of national income. Here the economy of a nation is modeled by considering four discrete functions: the national income y_n, the consumer expenditures C_n used to purchase goods, the investments I_n and the government expenditure G_n. The definition of y_n is

$$y_n = I_n + C_n + G_n. \tag{7.4.10}$$

The other relations among the four variables are based on the following assumptions, which are self-explanatory even from the economics point of view

$$C_{n+1} = \alpha y_n, \qquad 0 < \alpha < 1,$$

$$I_n = \rho(C_n - C_{n-1}), \qquad \rho > 0.$$

The parameter α is usually called marginal propensity to consume and ρ is the acceleration coefficients or "the relation." One usually assumes $G_n = G$ constant.

By substituting in (7.4.10) we get

$$y_n - \alpha(1 + \rho)y_{n-1} + \alpha\rho y_{n-2} - G = 0,$$

which is similar to (7.4.6). The constant solution (or equilibrium point) is now

$$y_e = \frac{G}{1 - \alpha}.$$

If this point is asymptotically stable, then $\lim_{n\to\infty} y_n = y_e$, which means that in this hypothesis the national income, under the effect of the government expenditure G will tend toward the income y_e, which is greater than G. The factor $\frac{1}{1-\alpha}$ is said to be a multiplier. To see when this happens, just repeat what is said for the previous model remembering that in this case ρ cannot be negative.

7.5. Models of Traffic in Channels

The models presented here concern the traffic in channels (for example telephone lines) and are from information and queueing theory. In the first simple model, the number of messages of given duration is derived. In the other two models the probability of requests for services (for example telephone calls) is derived as well as the probability of loss of the request in a system of limited channels.

Consider a channel and suppose that two elementary informations S_1 and S_2 of duration t_1 and t_2 respectively (for example the point and line in the Morse alphabet) can be combined in order to obtain a message. Let t be a time interval greater than both t_1 and t_2. One is interested in the number of messages N_t of length t. These messages can be divided in two classes: those ending with S_1 and those ending with S_2. The number of messages in the first class is N_{t-t_1} while the number of messages in the second class is N_{t-t_2}. Then we have

$$N_t = N_{t-t_1} + N_{t-t_2}. \tag{7.5.1}$$

Suppose, for simplicity, that $t_1 = 1$ and $t_2 = 2$. The previous formula becomes

$$N_t = N_{t-1} + N_{t-2}. \tag{7.5.2}$$

The initial conditions are $N_1 = 1$ and $N_2 = 2$. The equation (7.4.2) is the same that defines the Fibonacci sequence (see (2.3.16)).

The solution is

$$N_t = c_1\rho_1^t + c_2\rho_2^t,$$

where $\rho_1 = \dfrac{1+\sqrt{5}}{2}$, $\rho_2 = \dfrac{1-\sqrt{5}}{2}$. Because $|\rho_2| < 1$, for large t, one has $N_t \simeq c_1\rho_1^t$. Shannon defines the capacity of a channel as

$$c = \lim_{t\to\infty} \frac{\log_2 N_t}{t} = \log_2 \rho_1.$$

Difference equations arise very often in queuing theory. We shall give two examples. Consider a queue of individuals (or telephone calls in a channel). Let $P_n(t)$ be the probability of n items arrived at time t and we have $\sum_{n=0}^{\infty} P_n(t) = 1$.

Let $\lambda\Delta t$ be the probability that a single arrival during the small time interval Δt and suppose that the probability of more than one arrival in the same interval is negligible (Poisson process).

Let $\mu\Delta t$ be the probability of completing the service in the interval Δt. We shall assume that the service is a Poisson process, that is, the probability of no arrivals in Δt is $1 - \lambda\Delta t$ and the probability of the service not completed in Δt (no departures) is $1 - \lambda\Delta t$. The following model is due to Erlang (see [22]) and describes the situation where at the beginning there are no items in the single channel queue and the service is made on a first come, first served basis,

$$P_n(t+\Delta t) = P_n(t)(1-\lambda\Delta t)(1-\mu\Delta t) + P_{n-1}(t)\lambda\Delta t + P_{n+1}(t)\mu\Delta t,\ n \geq 1,$$

$$P_0(t+\Delta t) = P_0(t)(1-\lambda\Delta t) + P_1(t)\mu\Delta t.$$

The meaning of the equations is the following: the probability that at time $t+\Delta t$ there are n items in line is equal to the sum of three terms:

(1) The probability of already having n items at time t multipled by the probability of no arrivals during Δt and the probability of no departure in the same interval;

(2) the probability of having $n-1$ items at time t multiplied by the probability of a new arrival in Δt;
(3) the probability of having $n+1$ items at time t multiplied by the probability of a departure in Δt.

Taking the limit as $t \to 0$ one has

$$\frac{dP_n(t)}{dt} = -(\lambda + \mu)P_n(t) + \lambda P_{n-1}(t) + \mu P_{n+1}(t), \qquad n \geqq 1,$$

$$\frac{dP_0(t)}{dt} = -\lambda P_0(t) + \mu P_1(t).$$

It is important to know how the system behaves for large t. That is, to know if the limit $P_n = \lim_{t\to\infty} P_n(t)$ exists.

The probability of P_n will describe the steady state of the problem. In order to get this information one observes that in the steady state the derivatives will be zero.

It follows that the limiting probabilities P_n will satisfy the difference equations

$$\mu P_{n+1} - (\lambda + \mu)P_n + \lambda P_{n-1} = 0, \tag{7.5.3}$$

$$-\lambda P_0 + \mu P_1 = 0. \tag{7.5.4}$$

To this one must add the condition $\sum_{n=0}^{\infty} P_n = 1$, which simply states that it is certain that in the system we must have either no items, or more items. Let $\rho = \dfrac{\lambda}{\mu}$. The solution of (7.5.3) is given in terms of the roots of the polynomial equation

$$z^2 - (1+\rho)z + \rho = 0,$$

which are $z_1 = 1$, $z_2 = \rho$. If $\rho < 1$, we get

$$P_n = c_1 + c_2\rho^n.$$

Considering that from (7.5.4) one has $P_1 = \rho P_0$, it follows that $c_1 = 0$. To derive c_2 one uses the additional condition

$$1 = \sum_{n=0}^{\infty} P_n = c_2 \frac{1}{1-\rho},$$

which gives $c_2 = 1 - \rho$ and then $P_n = (1-\rho)\rho^n$, which is called geometric distribution.

Important statistical parameters are:

(1) the expected number in the system

$$L = \sum_{n=0}^{n} nP_n = (1-\rho)\sum_{n=0}^{\infty} n\rho^n = (1-\rho)\rho\frac{d}{d\rho}\sum_{n=0}^{\infty}\rho^n = \frac{\rho}{1-\rho};$$

(2) the variance

$$V = \sum_{n=0}^{\infty}(n-L)^2 P_n = \sum_{n=0}^{\infty} n^2 P_n - L^2,$$

one shows that $\sum_{n=0}^{\infty} n^2 P_n = L + 2L^2$ and $V = L + L^2$;

(3) the expected number in the line

$$L_q = \sum_{n=0}^{\infty}(n-1)P_n = \rho L.$$

In the following generalization one considers the case of N identical channels with the same hypothesis on the distribution of arrivals and departures (Poisson distribution). Moreover, we shall assume that both the state with zero items E_0, and the state with N items E_N are reflecting, that is, from these states transactions are possible only in one direction. From the state E_0 a possible transaction will be to the state E_1 and from the state E_N a possible transaction will be to the state E_{N-1}. The resulting model is

$$\frac{dP_n(t)}{n} = -(\lambda + n\mu)P_n(t) + \lambda P_{n-1}(t) + (n+1)\mu P_{n+1}(t),\ 1 \le n \le N-1,$$

$$\frac{dP_0}{dt} = -\lambda P_0(t) + \mu P_1(t),$$

$$\frac{dP_N}{dt} = -N\mu P_N(t) + \lambda P_{N-1}(t).$$

The steady state solution satisfies the difference equations

$$(n+1)P_{n+1} - (\rho + n)P_n + \rho P_{n-1} = 0,$$

$$P_1 = \rho P_0,$$

$$P_N = \frac{\rho}{N} P_{N-1},$$

where, as before $\rho = \dfrac{\lambda}{\mu}$. Let us look for solutions of the form $P_n = z^n\rho^n$, from which one has $z_0 = P_0$, $z_1 = P_0$, $z_N = \dfrac{z_{N-1}}{N}$. The equation in is then

$$\rho^n[(-z_n + (n+1)z_{n+1})\rho - (-z_{n-1} + nz_n)] = 0.$$

Let $\theta_n = -z_n + (n+1)z_{n+1}$. The equation becomes

$$\rho\theta_n - \theta_{n-1} = 0,$$

whose solution is

$$\theta_n = \rho^{-n}\theta_0.$$

For $n = N - 1$, we must have $\theta_N = -z_{N-1} + Nz_N = -z_{N-1} + z_{N-1} = 0$. It follows then $\theta_0 = 0$. The equation for z_n is then

$$-z_n + (n+1)z_{n+1} = 0,$$

which has the solution $z_n = \dfrac{1}{n!} z_0 = \dfrac{1}{n!} P_0$. Going back to P_n we have $P_n = \dfrac{1}{n!}\rho^n P_0$. Imposing the condition $\sum_{n=0}^{N} P_n = 1$, one finds $P_0 = 1/\sum_{j=0}^{N} \dfrac{\rho^j}{j!}$. The solution is then

$$P_n = \frac{\dfrac{\rho^n}{n!}}{\sum_{j=0}^{N} \dfrac{\rho^j}{j!}},$$

which is called Erlang's loss distribution. For $n = N$, one obtains the probability that all the channels are busy, which also gives the probability that a call is lost. As $N \to \infty$ one obtains the Poisson distribution $P_n = \dfrac{\rho^n}{n!} e^{-\rho}$.

The previous model also describes the parking lot problem. For this problem, N is the number of places or slots in the parking lot. The reflecting condition at the end just describes the policy that the parking lot closes when it is full. The probability that a car cannot be parked is P_N. If the management follows the policy of allowing a queue waiting for a free place, then the model is modified as follows:

$$P_1 = \rho P_0,$$

$$(n+1)P_{n+1} - (\rho + n)P_n + \rho P_{n-1} = 0, \qquad n \leq N,$$

$$NP_{N+1} - (\rho + N)P_n + \rho P_{n-1} = 0, \qquad n > N.$$

The solution is

$$P_n = \frac{\rho^n}{n!} P_0, \qquad n \leq N,$$

as before, and

$$P_n = \frac{\rho^n}{N!N^{n-N}} P_0, \qquad n > N.$$

Using the relation $\sum_{n=0}^{\infty} P_n = 1$ one obtains P_0.

7.6. Problems

7.1. In the case $\alpha_i > 0$, derive the existence of eigenvalue of greatest absolute value from Theorem 7.1.1 instead of Perron–Frobenius theorem.

7.2. Write the generating function for the solution of (7.1.8) and derive the behavior of the solution (Renewal Theorem).

7.3. Show that for $a < 1$ the origin is asymptotically stable for $y_0 \in (0, 1)$ for the discrete logistic equation. (Hint: use the Lyapunov function $V_n = y_n^2$.)

7.4. Discuss the stability of $\bar{y}$ for the logistic equation using $V_n = (y_n - \bar{y})^2$.

7.5. Determine the solution of (7.1.5) for $M = 4$ and $a_1 = a_2 = a_3 = 0$, $a_4 = 16$.

7.6. Discuss the modified Cobweb model with $\rho = -1$.

7.7. Derive the national income model assuming that $I_n = \rho(y_{n-1} - y_{n-2})$.

7.8. Suppose that the input for the traffic in N channels has Poisson distribution and depends on the numbers of free channels, that is, ρ_i is not constant. Derive the model and find the solution.

7.7. Notes

Age dependent population models have many names associated with them such as Lotka, McKendrick, Bernardelli and Leslie. An extensive treatment of the subject may be found in Hoppensteadt [80], Svirezhev and Logofet [164]. In [164] we also have several references to the subject. Theorem 7.1.1, which is a generalization of a Cauchy theorem, may be found in Ostrowsky [121], see also Pollard [141].

The discrete logistic equation has been discussed by many authors, for example, Lorenz [97], May [106, 107], Hoppensteadt [80], Hoppensteadt and Hyman [79]. The result of Li and Yorke [95] on the existence of periodic solutions is contained in a more general result due to Sharkovsky, see [155], [161], [79], [162] and [8] for more recent proofs and results. When a difference equation is derived from an approximation of a continuous differential equation, the question arises to what extent the behavior of the respective solutions are similar. This problem has been outlined and treated by Yamaguti et al [176, 177, 178] and Potts [145]. The distillation model has been adapted from [110], see also [173]. The discrete models in economics may be found in Gandolfo [50], Goldberg [60] and Luenberger [98]. Queueing theory is a large source of difference equations, see for example Saaty [149], where one may find very interesting traffic models more general than those presented in Section 7.4. For the parking lot problem, as well as other models concerning traffic flow, see Height [66].

Appendix A

A.1. Function of Matrices

Let A be a $\bar{n} \times \bar{n}$ matrix with real or complex elements, and let us consider expression defined by

$$P(A) = \sum_{i=0}^{k} p_i A^i, \tag{A.1.1}$$

where $k \in \mathbf{N}^+$ and $p_i \in \mathbb{C}$. Such expressions are said to be matrix polynomials. Associated with them are the polynomials $p(z) = \sum_{i=0}^{k} p_i z^i$.

There are some differences of algebraic nature among polynomials defined in the complex field and those defined on a commutative ring (the set of powers of a matrix). For example, it can happen that $A^n = AA^{n-1} = 0$ but neither A or A^{n-1} is the null matrix. More generally, it can happen that $p(A) = \prod_{i=1}^{k} (A - z_i I)$ with $z_i \in \mathbb{C}$ is the null matrix and $A \neq z_i I$, which cannot happen for complex polynomials.

The Cayley theorem states that if $z_1, z_2, \ldots, z_n$ are the eigenvalues of A, with multiplicity $\bar{m}_i$, letting

$$p(A) = \prod_{i=1}^{s} (A - z_i I)^{\bar{m}_i} \tag{A.1.2}$$

one has $p(A) = 0$. The polynomial $p(z)$ is called the characteristic polynomial.

Let $\psi(z)$ be the monic polynomial of minimal degree such that

$$\psi(A) = 0, \tag{A.1.3}$$

and let n be its degree.

Theorem A.1.1. *Every root of the minimal polynomial is also a root of the characteristic polynomial and vice versa.*

Proof. Dividing $p(z)$ by $\psi(z)$ we find two polynomials $q(z)$ and $r(z)$ such that

$$p(z) = q(z)\psi(z) + r(z) \tag{A.1.4}$$

with degree of $r(z)$ less than n. For the corresponding matrix polynomials we have $0 = p(A) = q(A)\psi(A) + r(A)$. Since $\psi(A) = 0$ and it is minimal, it follows that $r(A)$ is identically zero. Thus (A.1.4) becomes $p(z) = q(z)\psi(z)$ proving that the roots of $\psi(z)$ are necessarily roots of $p(z)$. Suppose now that $\lambda \in \mathbb{C}$ is a root of $p(z)$ and therefore it is also an eigenvalue of A. For every $j \in \mathbf{N}^+$, it follows that $A^j x = \lambda^j x$, where x is the corresponding eigenvector. Then we have $\psi(A)x = \psi(\lambda)x$ from which it follows $\psi(\lambda) = 0$, proving that λ is also a root of $\psi(z)$. ■

Let $h(z)$ be any other polynomial of degree greater than n. We can write $h(z) = q(z)\psi(z) + r(z)$ where the degree of $r(z)$ is less than n. Consequently

$$h(A) = r(A). \tag{A.1.5}$$

The last result shows that a polynomial of degree greater than n and a polynomial of degree less than n may represent the same matrix.

Let $h(z)$ and $g(z)$ be two polynomials such that $h(A) = g(A)$. Then the polynomial

$$d(z) = h(z) - g(z) \tag{A.1.6}$$

annihilates A; that is, $d(A) = 0$. It follows that it is possible to find $q(z)$ such that $d(z) = q(z)\psi(z)$. Let $m_1, m_2, \ldots, m_s$ be the multiplicities of the roots of $\psi(z)$. Then we have $\psi(z_i) = \psi'(z_i) = \ldots = \psi^{(m_i-1)}(z_i) = 0$, $i = 1, 2, \ldots, s$, from which $d(z_i) = d'(z_i) = \ldots = d^{(m_i-1)}(z_i) = 0$, $i = 1, 2, \ldots, s$. This result shows that

$$\begin{cases} h(z_i) = g(z_i), \\ h'(z_i) = g'(z_i), \\ \vdots \\ h^{(m_i-1)}(z_i) = g^{(m_i-1)}(z_i), \end{cases} \tag{A.1.7}$$

for $i = 1, 2, \ldots, s$. Two polynomials satisfying the above condition are said to assume the same values on the spectrum of A. Now suppose that the two polynomials assume the same values on the spectrum of A. It follows that $d(z)$ has multiple roots z_i of multiplicity m_i and then it will be divided exactly by $\psi(z)$; that is, $d(z) = q(z)\psi(z)$. Then $d(A) = g(A) - h(A) = 0$ and the two polynomials represent the same matrix.

The foregoing arguments prove the following result.

Theorem A.1.2. *Let $g(z)$ and $h(z)$ be two polynomials on $\mathbb{C}$ and $A \in C$. Then $g(A) = h(A)$ iff they assume the same values on the spectrum of A.*

Definition A.1.1. Let $f(z)$ be a complex valued function defined on the spectrum of A, and let $g(z)$ be the polynomial assuming the same values on the spectrum of A. The matrix function $f(A)$ is defined by $f(A) = g(A)$.

The problem of defining $f(A)$ is then solved once we find the polynomial $g(z)$ such that $g^{(i)}(z_k) = f^{(i)}(z_k)$ for $k = 1, 2, \ldots, s$, $i = 0, 1, \ldots, m_{k-1}$. The last problem is solved by constructing the interpolating polynomial (Lagrange–Hermite polynomial)

$$g(z) = \sum_{k=1}^{s} \sum_{i=1}^{m_k} f^{(i-1)}(z_k)\phi_{ki}(z), \tag{A.1.8}$$

where $\phi_{ki}(z)$ are polynomials of degree $n - 1$ such that

$$\phi_{ki}^{(r-1)}(z_j) = \delta_{kj}\delta_{ri}, \qquad \begin{cases} j, k = 1, 2, \ldots, s, \\ i, r = 1, 2, \ldots, m_k. \end{cases}$$

For example for $m_i = 1$, $i = 1, 2, \ldots, n$,

$$\phi_{k1}(z) = \frac{\prod_{j \neq k} (z - z_j)}{\prod_{j \neq k} (z_k - z_j)}.$$

It can be proved that the functions $\phi_{ki}(z)$ are linearly independent.

The matrix polynomial is then

$$g(A) = \sum_{k=1}^{s} \sum_{i=1}^{m_k} f^{(i-1)}(z_k)\phi_{ki}(A). \tag{A.1.9}$$

The matrices $\phi_{ki}(A)$ are said to be components of A. They are independent of the function $f(z)$. As usual in this field we shall put

$$\phi_{ki}(A) = Z_{ki}. \tag{A.1.10}$$

These matrices, being polynomials of A, commute with A. With this notation (A.1.10) becomes

$$f(A) \doteq g(A) = \sum_{k=1}^{s} \sum_{i=1}^{m_k} f^{(i-1)}(z_k)Z_{ki}. \tag{A.1.11}$$

The set of matrices Z_{ki} are linearly independent. In fact if there exist constants c_{ki} not all zero such that $\sum_{k=1}^{s} \sum_{i=1}^{m_k} c_{ki}Z_{ki} = 0$, the associated polynomial $s(z) = \sum_{k=1}^{s} \sum_{i=1}^{m_k} c_{ki}\phi_{ki}(z)$ would annihilate A.

A.2. Properties of Component Matrices and Sequences of Matrices

Let us list some properties of the component matrices Z_{ki}.

Taking $f(z) = 1$, we get from (A.1.11)

$$\sum_{k=1}^{s} Z_{k1} = I. \tag{A.2.1}$$

Also, taking $f(z) = z$, we see that

$$A = \sum_{k=1}^{s} z_k Z_{k1} + \sum_{k=1}^{s} Z_{k2}, \tag{A.2.2}$$

from which

$$A^2 = \sum_{k=1}^{s} (z_k Z_{k1} + Z_{k2}) \sum_{j=1}^{s} (z_j Z_{j1} + Z_{j2})$$

$$= \sum_{k=1}^{s} \sum_{j=1}^{s} (z_k z_j Z_{k1} Z_{j1} + 2 z_k Z_{k1} Z_{j2} + Z_{k2} Z_{j2}).$$

Starting directly from $f(z) = z^2$, one has

$$A^2 = \sum_{k=1}^{s} (z_k^2 Z_{k1} + 2 z_k Z_{k2} + 2 Z_{k3}).$$

Comparing the two results, it follows that

$$Z_{k1} Z_{j1} = \delta_{kj} Z_{k1},$$

$$Z_{k1} Z_{j2} = \delta_{kj} Z_{k2},$$

$$Z_{k2} Z_{j2} = 2\delta_{kj} Z_{k3}.$$

Proceeding in a similar way, it can be proved in general that

$$\begin{aligned} Z_{kp} Z_{ir} &= 0 && \text{if } k \neq i, \\ Z_{kp} Z_{k1} &= \mathbf{Z}_{kp} && p \geq 1, \\ Z_{kp} Z_{k2} &= p\mathbf{Z}_{k,p+1} && p \geq 2. \end{aligned} \tag{A.2.3}$$

From the last relation it follows easily that

$$Z_{kp} = \frac{1}{(p-1)!} Z_{k2}^{p-1}, \qquad p \geq 2. \tag{A.2.4}$$

It is worth noting that, from the second relation of (A.2.3), we get

$$Z_{k1}^2 = Z_{k1}, \tag{A.2.5}$$

showing that the matrices Z_{k1} are projections. Multiplying the expression $A - z_i I = \sum_{k=1}^{s} (z_k - z_i)Z_{k1} + \sum_{k=1}^{s} Z_{k2}$, by Z_{i1} and considering (A.2.3), one gets,

$$(A - z_i I)Z_{i1} = \sum_{k=1}^{s} (z_k - z_i)Z_{k1}Z_{i1} + \sum_{k=1}^{s} Z_{k2}Z_{i1} = Z_{i2}. \tag{A.2.6}$$

Because of (A.2.4) and (A.2.6), we obtain

$$Z_{kp} = \frac{1}{(p-1)!} Z_{k2}^{p-1} = \frac{1}{(p-1)!} (A - z_k I)^{p-1} Z_{k1}. \tag{A.2.7}$$

From this result, it follows that (A.1.11) can be written as

$$f(A) = \sum_{k=1}^{s} \sum_{i=1}^{m_k} \frac{f^{(i-1)}(z_k)}{(i-1)!} (A - z_k I)^{i-1} Z_{k1}. \tag{A.2.8}$$

Let us consider now the function $f(z) = \dfrac{1}{z - \lambda}$, where $\lambda \neq z_k$. The function $f(A) = (A - \lambda I)^{-1}$ is expressed by

$$(A - \lambda I)^{-1} = -\sum_{k=1}^{s} \sum_{i=1}^{m_k} (i-1)!(\lambda - z_k)^{-i} Z_{ki} \tag{A.2.9}$$

from which, considering (A.2.4), we get

$$\begin{aligned}(A - \lambda I)^{-1} = -\sum_{k=1}^{s} [&(\lambda - z_k)^{-1} Z_{k1} + (\lambda - z_k)^{-2} Z_{k2} \\ &+ (\lambda - z_k)^{-3} Z_{k2}^2 + \ldots + (\lambda - z_k)^{-m_k} Z_{k2}^{m_{k-1}}],\end{aligned}$$

we then obtain because of $(A - \lambda I) = -\sum_{k=1}^{s} [(\lambda - z_k)Z_{k1} - Z_{k2}]$,

$$\begin{aligned} I = \sum_{k=1}^{s} [&Z_{k1} + (\lambda - z_k)^{-1} Z_{k1}Z_{k2} + (\lambda - z_k)^{-2} Z_{k1}Z_{k2}^2 + \ldots \\ &+ (\lambda - a_k)^{-m_k+1} Z_{k1}Z_{k2}^{m_k-1} \\ &- (\lambda - z_k)^{-1} Z_{k1}Z_{k2} - (\lambda - z_k)^{-2} Z_{k2}^2 - (\lambda - z_k)^{-3} Z_{k2}^3 - \ldots \\ &- (\lambda - z_k)^{-m_k} Z_{k2}^{m_k}] \\ = \sum_{k=1}^{s} [&Z_{k1} - (\lambda - z_k)^{-m_k} Z_{k2}^{m_k}], \end{aligned}$$

from which results

$$Z_{k2}^{m_k} = 0, \tag{A.2.10}$$

showing that the matrices Z_{k2} are nilpotent.

This result allows us to extend the internal sum in formula (A.2.8) up to $\bar{m}_k$, the multiplicity of the characteristic polynomial. In fact, since $m_k \le \bar{m}_k$, $Z_{k2}^{\bar{m}_k+j} = 0$ for $j = 0, 1, \ldots, \bar{m}_k - m_k$, and hence (A.2.8) can be written as

$$f(A) = \sum_{k=1}^{s} \sum_{i=1}^{\bar{m}_k} \frac{f^{(i-1)}(z_k)}{(i-1)!}(A - z_k I)^{i-1} Z_{k1}. \tag{A.2.11}$$

Now consider a sequence of complex valued functions $f_1(z), f_2(z), \ldots$ defined on the spectrum of A.

Definition A.2.1. The sequence $f_1, f_2, \ldots,$ converges on the spectrum of A if, for $k = 1, 2, \ldots, s$, $\lim_{i\to\infty} f_i(z_k) = f(z_k)$, $\lim_{i\to\infty} f_i'(z_k) = f'(z_k)$, and so on until $\lim_{i\to\infty} f_i^{(m_j-1)}(z_k) = f^{(m_j-1)}(z_k)$. The following theorem is almost evident (see Lancaster).

Theorem A.2.1. *A sequence of matrices $f_i(A)$ converges if the sequence $f_i(z)$ converges on the spectrum of A.*

Corollary A.2.1. *Let A be an $s \times s$ complex matrix having all the eigenvalues inside the complex unit disk. Then $(I - A)^{-1} = \sum_{i=0}^{\infty} A^i$.*

Proof. Consider for $i \ge 0$ the sequence $f_i(z) = \sum_{j=0}^{i} z^j$. This sequence (and the sequences obtained by differentiation) converges if $|z| < 1$ to $(1 - z)^{-1}$. It follows that there exists the limit of $f_i(A)$ and the limit is $(I - A)^{-1}$. ■

Corollary A.2.2. *Let A be an $s \times s$ complex matrix. Then $e^A = \sum_{j=0}^{\infty} \frac{A^j}{j!}$.*

Proof. Consider for $i > 0$, the sequence $f_i(z) = \sum_{j=0}^{i} \frac{z^j}{j!}$. This sequence converges for all $z \in \mathbb{C}$. Likewise for $f_i^{(s)}(z)$. It follows that $f_i(A)$ converges to e^A. ■

Definition A.2.2. An eigenvalue z_k is said simple if $\bar{m}_k = 1$.

Definition A.2.3. An eigenvalue z_k is said semisimple if $Z_{k2} = 0$. A semisimple eigenvalue is not simple (or degenerate) if $m_k = 1$ and $\bar{m}_k \ne 1$.

In both cases the terms containing $(A - z_k I)^{p-1} Z_{k1}$ with $p > 2$ are not present in the expression of $f(A)$.

Example A.2.1. An important class of matrices are the so called companion matrices, defined by

$$A = \begin{pmatrix} 0 & 1 & 0 & \dots & 0 \\ 0 & 0 & 1 & & \vdots \\ \vdots & & & & 0 \\ 0 & & \dots & 0 & 1 \\ -\alpha_0 & -\alpha_1 & & \dots & -\alpha_{n-1} \end{pmatrix}.$$

For this class of matrices $m_i = \bar{m}_i$ (for all values of i), that is, the characteristic polynomial and the minimal polynomial coincide (see Lancaster) and this implies that there are no semisimple roots.

Example A.2.2. Let $f(z) = e^{zt}$ and A be an $n \times n$ matrix. Then

$$e^{At} = \sum_{k=1}^{s} \sum_{i=1}^{\bar{m}_k} \frac{t^{i-1}}{(i-1)!} e^{z_k t}(A - z_k I)^{i-1} Z_{k1}. \tag{A.2.12}$$

If the eigenvalues are all simple, the previous relation becomes

$$e^{At} = \sum_{k=1}^{\bar{n}} e^{z_k t} Z_{k1}. \tag{A.2.13}$$

If z_j is a semisimple eigenvalue, (A.2.11) becomes

$$e^{At} = \sum_{\substack{k=1 \\ k \neq j}}^{s} \sum_{i=1}^{\bar{m}_k} \frac{t^{i-1}}{(i-1)!} e^{z_k t}(A - z_k t)^{i-1} Z_{k1} + e^{z_j t} Z_{j1}. \tag{A.2.14}$$

Theorem A.2.2. *If* $\mathbf{Re} z_i < 0$ *for* $i = 1, 2, \dots, \bar{n}$, *then* $\lim_{t \to \infty} l^{At} = 0$.

Proof. From (A.2.12) it follows easily. ∎

Theorem A.2.3. *If for* $i = 1, 2, \dots, \bar{n}$, $\mathbf{Re} z_i \leq 0$ *and those for which* $\mathbf{Re} z_i = 0$ *are semisimple eigenvalues, then* e^{At} *tends to a bounded matrix as* $t \to \infty$.

Proof. It follows easily from (A.2.13). ∎

Example A.2.3. Let $f(z) = z^n$ and

$$\begin{aligned} A^n &= \sum_{k=1}^{s} \sum_{i=1}^{\bar{m}_k} \frac{n^{(i-1)}}{(i-1)!} z_k^{n-i+1} (A - z_k I)^{i-1} Z_{k1} \\ &= \sum_{k=1}^{s} \sum_{i=0}^{\bar{m}_k - 1} \binom{n}{i} z_k^{n-i} (A - z_k I)^i Z_{k1}. \end{aligned}$$

If the eigenvalues z_j of A are all simple, then (A.2.15) becomes

$$A^n = \sum_{k=1}^{\bar{n}} z_k^n Z_{k1}. \tag{A.2.16}$$

If z_j is semisimple, (A.2.14) reduces to

$$A^n = \sum_{\substack{k=1 \\ k \neq j}}^{s} \sum_{i=0}^{m_k - 1} \binom{n}{i} z_k^{n-i} (A - z_k I)^i Z_{k1} + z_j^n Z_{j1}. \tag{A.2.17}$$

Theorem A.2.4. *If* $|z_i| < 1$ *for* $i = 1, 2, \ldots, \bar{n}$, *then* $\lim_{n \to \infty} A^n = 0$.

Proof. The proof is easy to see from (A.2.15). ■

Theorem A.2.5. *If for* $i = 1, 2, \ldots, \bar{n}$, $|z_i| \leq 1$ *and the eigenvalues for which* $|z_i| = 1$ *are semisimple, then* A^n *tends to a bounded matrix as* $n \to \infty$.

Proof. The proof is easy to see from (A.2.17). ■

It may happen, however, that for multiplicity of higher order and consequently higher dimension of the matrix, the terms in (A.2.17) may become large. For example, let us take the matrix

$$A = \begin{pmatrix} \lambda & \beta & 0 & & \cdots & 0 \\ 0 & \lambda & \beta & 0 & \cdots & 0 \\ \vdots & & & & \ddots & \\ & & & & & 0 \\ 0 & & \cdots & & 0 & \lambda \end{pmatrix}_{s \times s} = \lambda I + \beta H \tag{A.2.18}$$

with

$$H = \begin{pmatrix} 0 & 1 & 0 & & \cdots & 0 \\ 0 & 0 & 1 & 0 & \cdots & 0 \\ & & & & \ddots & \\ \vdots & & & & \ddots & 0 \\ & & & & \ddots & 1 \\ 0 & & \cdots & & & 0 \end{pmatrix}_{s \times s} \tag{A.2.19}$$

and $H^s = 0$. Then (A.2.17) becomes

$$A^n = \sum_{i=0}^{s-1} \binom{n}{i} \lambda^{n-i} \beta^i H^i. \tag{A.2.20}$$

Multiplying by $E = (1, 1, \ldots, 1)^T$ and taking $n = s - 1$, we have

$$A^n E = \sum_{i=1}^{n} \binom{n}{i} \lambda^{n-i} \beta^i H^i E. \tag{A.2.21}$$

The first component of this vector is

$$(A^n E)_1 = \sum_{i=0}^{n} \binom{n}{i} \lambda^{n-i} \beta^i = (\lambda + \beta)^n, \tag{A.2.22}$$

from which it is seen that

$$\|A^n E\| \geq |\lambda + \beta|^n. \tag{A.2.23}$$

This implies that the component of $A^n E$ will grow even if $|\lambda| < 1$, but $|\lambda + \beta| > 1$. Eventually they will tend to zero for $n > s - 1$, because from (A.2.20) the exponents of λ become bigger and bigger while those of β remain bounded by $s - 1$. But in the cases (as in the matrix arising in the discretization of P.D.E.), where s grows itself, the previous example shows that the eigenvalues are not enough to describe the behavior of A^n. This will lead to the introduction of the concept of spectrum of a family of matrices.

A.3. Integral Form of a Function of Matrix

Let $M(z)$ be an $s \times s$ matrix whose components $m_{ij}(z)$ are functions of the variable z. In the following, we shall assume that the functions $m_{ij}(z)$ are analytic functions in some specified domain of the complex plane.

Definition A.3.1. The derivative of the matrix $M(z)$ with respect to z, is the matrix $M'(z)$ whose elements are $m'_{ij}(z)$.

Definition A.3.2. The integral of the matrix $M(z)$ is the matrix whose coefficients are the integrals of $m_{ij}(z)$.

Let us consider now the matrix

$$\boldsymbol{R}(z, A) \equiv (zI - A)^{-1}, \tag{A.3.1}$$

called resolvent matrix. From (A.2.9), we get $\boldsymbol{R}(z, A) = \sum_{k=1}^{s} \sum_{i=1}^{m_k} (i-1)!(z - z_k)^{-i} Z_{ki}$, which has singularities for $z = z_k$, ($k = 1, 2, \ldots, s$). Suppose now that Γ_h is a circle around z_k not containing other eigenvalues. Then (A.2.9) becomes

$$\begin{aligned} \boldsymbol{R}(z, A) &= \sum_{j=1}^{m_h} (j-1)!(z - z_h)^{-j} Z_{hj} + \sum_{\substack{k=1 \\ k \neq h}}^{s} \sum_{j=1}^{m_k} (j-1)!(z - z_k)^{-j} Z_{kj} \\ &= \sum_{j=1}^{m_h} (j-1)!(z - z_h)^{-j} Z_{hj} + S_h(z), \end{aligned}$$

where $S_h(z)$ is the matrix obtained by expanding the term $(z - z_k)^{-1} = \frac{1}{z_h - z_k}\left(1 - \frac{z + z_h}{z_h - z_k}\right)^{-1}$. It is seen that the elements of $S_h(z)$ are analytic in Γ_h. Integrating on Γ_h, we have

$$\int_{\Gamma_h} R(z, A)\, dz = \sum_{j=1}^{m_h} (j-1)!\, Z_{hj} \int_{\Gamma_h} (z - z_h)^{-j}\, dz = 2\pi i Z_{h1},$$

from which

$$Z_{h1} = \frac{1}{2\pi i} \int_{\Gamma_h} \boldsymbol{R}(z, A)\, dz. \tag{A.3.3}$$

Analogously, for $t = 1, 2, \ldots, m_h - 1$,

$$\begin{aligned} \int_{\Gamma_h} (z - z_h)^t R(z, A)\, dz &= \sum_{j=1}^{m_h - 1} (j-1)!\, Z_{hj} \int_{\Gamma_h} \frac{(z - z_h)^t}{(z - z_h)^j}\, dz \\ &= (2\pi i)\, t!\, Z_{h,t+1}. \end{aligned} \tag{A.3.4}$$

In general we shall obtain

$$Z_{hj} = \frac{1}{(j-1)!\, 2\pi i} \int_{\Gamma_h} (z - z_h)^{j-1} \boldsymbol{R}(z, A)\, dz. \tag{A.3.5}$$

Theorem A.3.1. *Suppose that f is a function continuous on a closed and regular curve Γ, which contains in its interior the points $z_1, z_2, \ldots, z_s$ and that f is analytic inside. Then*

$$f(A) = \frac{1}{2\pi i} \int_{\Gamma} f(z) \boldsymbol{R}(z, A)\, dz. \tag{A.3.6}$$

Proof. From (A.2.9),

$$\begin{aligned} \frac{1}{2\pi i} \int_{\Gamma} f(z) R(z, A)\, dz &= \sum_k \sum_j (j-1)!\, Z_{kj} \frac{1}{2\pi i} \int_{\Gamma} (z - z_k)^{-j} f(z)\, dz \\ &= \sum_k \sum_j (j-1)!\, Z_{kj} \frac{f^{(j-1)}(z)}{(j-1)!} = f(A), \end{aligned}$$

where the Cauchy formula has been applied. ∎

A.4. Jordan Canonical Form

From (A.2.2), it results that

$$A = \sum_{k=1}^{s} (z_k Z_{k1} + Z_{k2}) \tag{A.4.1}$$

and from (A.2.1) it follows that the space $\boldsymbol{R}^n$ can be decomposed into s subspaces $M_1, \ldots, M_s$ defined by

$$M_j = Z_{j1}\boldsymbol{R}^n, \qquad j = 1, 2, \ldots, s. \tag{A.4.2}$$

Similarly, from (A.2.3) it follows that for $i \neq j$

$$M_i \cap M_j = 0. \tag{A.4.3}$$

Let $\bar{m}_i = \dim M_i$. Then $\sum_{i=1}^{s} \bar{m}_i = n$ and it is seen that $\bar{m}_i$ is the multiplicity of z_i as root of the characteristic polynomial. We want to choose appropriately a base for the subspace M_j.

Lemma A.4.1. *Let $x^{(j)} \in M_j$, $Z_{j2}^i x^{(j)} \neq 0$ for $0 \le i < p$ and $Z_{j2}^p x^{(j)} = 0$. Then the vectors $Z_{j2}^i x^{(j)}$, $i = 0, 1, \ldots, p-1$ are linearly independent.*

Proof. From

$$\sum_{i=0}^{p-1} c_i Z_{j2}^i x^{(j)} = 0, \tag{A.4.4}$$

it follows, multiplying successively by $Z_{j2}^{p-1}, Z_{j2}^{p-2}, \ldots$, that $c_i = 0$. We shall call the set of vectors $Z_{j2}^i x^{(j)}$ for $i = 1, 2, \ldots, p-1$ the chain starting at $x^{(j)}$. These vectors are also called generalized eigenvectors associated to $x^{(j)}$. ■

Lemma A.4.2. *Let $x^{(j)} \in M_j$, $Z_{j2}^i x^{(j)} \neq 0$ for $0 \le i < p$ and $Z_{j2}^p x^{(j)} = 0$. Suppose that there exists $y^{(j)} \in M_j$ linearly independent from the previous defined set of vectors $\{Z_{j2}^i z^{(j)}\}$. Then the vectors $\{Z_{j2}^i y^{(j)}\}$ are linearly independent from the previous set.*

Proof. Similar to the previous proof. ■

Going back to our problem we shall distinguish the following cases:

(1) If $m_i = 1 = \bar{m}_1$ for $i = 1, 2, \ldots, n$, it follows from (A.4.1), $x_i \in M_i$, that

$$Ax_i = z_i x_i. \tag{A.4.5}$$

Defining the matrix $X = (x_1, \ldots, x_n)$, we obtain

$$AX = X\begin{pmatrix} z_1 & 0 & \ldots & 0 \\ 0 & & & \vdots \\ \vdots & & & 0 \\ 0 & & 0 & z_n \end{pmatrix} \tag{A.4.6}$$

and considering that the matrix X is nonsingular we define the diagonal matrix B similar to A by

$$B = X^{-1}AX. \tag{A.4.7}$$

(2) $m_i = 1$, $m_i < \bar{m}_i$. In this case we can choose $\bar{m}_i$ linearly independent vectors in M_i, such that (it is $Z_{i2} = 0$):

$$Ax_j = z_i x_j, \qquad j = 1, 2, \ldots, \bar{m}_i$$

and the matrix A can be transformed in diagonal form, as in the previous case.

(3) At least an m_j is different from 1, $m_j = \bar{m}_j$ and $Z^t_{j2} \neq 0$ for $t\_m_j - 1$. Consider the choice of vectors $x^{(j)}_i = Z^i_{j2}x^{(j)}_0$, $i = 0, 1, \ldots, m_j - 1$, where $x^{(j)}_0$ is a vector in M_j. From Lemma A.4.1, these vectors are linearly independent and can be chosen as a base for M_j.

From (A.4.1) we have

$$Ax^{(j)}_i = z_i x^{(j)}_i + Z_{k2}x^{(j)}_i = z_i x^{(j)}_i + x^{(j)}_{i+1}, \qquad i = 0, 1, \ldots, m_j - 2,$$

$$Ax^{(j)}_{m_j-1} = z_j x^{(j)}_{m_j-1}.$$

Defining the matrix $X = (x^{(1)}_0, \ldots, x^{(1)}_{m_j-1}, \ldots, x^{(s)}_0, \ldots, x^{(s)}_{m_s-1})$, we have from (A.4.8)

$$AX = X\begin{pmatrix} z_1 & & 1 & & & & \\ & \ddots & & \ddots & & & \\ & & z_1 & & 0 & & \\ & & & z_2 & & \ddots & \\ & & & & \ddots & & \\ & & & & z_2 & & 1 \\ & & & & & \ddots & & \ddots \end{pmatrix}.$$

Since X is nonsingular, a matrix J, similar to A can be defined as

$$J = X^{-1}AX = \mathrm{diag}(J_1, J_2, \ldots, J_s).$$

The structure of B is a block diagonal, each block has the form

$$J_k = \begin{pmatrix} \lambda_k & 1 & 0 \\ 0 & \ddots & \ddots \\ \vdots & \ddots & 1 \\ 0 & 0 & \lambda_k \end{pmatrix}. \tag{A.4.9}$$

(4) At least an m_j is different from 1 and $Z^t_{j2} \neq 0$ for $0 \leq t < m_j$. Choosing $x^{(j)}_0 \in M_j$, we define the set of vectors $x^{(j)}_i = Z^i_{j2}x^{(j)}_0$ for $i = 0, 1, \ldots, \bar{t} \leq t$. These $\bar{t}$ vectors are linearly independent (see Lemma

1), but they are not enough to form a base of M_j. Choose another vector $x_{\bar{i}}^{(j)}$ independent of the previous ones and define $x_{\bar{i}+1}^{(j)} = Z_{j2}x_{\bar{i}}^{(j)}$ and so on until a set of $\bar{m}_j$ linearly independent vectors has been found. Then we proceed as in the previous case.

The matrix J is block diagonal as in the previous case, but the block corresponding to the subspaces M_j can be decomposed in different subblocks with each one corresponding to one chain of vectors. Each subblock is of the type (A.4.9). The matrix J is said to be the Jordan canonical form of the matrix A.

Each chain associated to M_j contains an eigenvector associated to z_j (the last vector of the chain). The number of eigenvectors associated to z_j is the geometric multiplicity of the chain. The dimension of M_j is the algebraic multiplicity of z_j.

A.5. Norms of Matrices and Related Topics

The definitions of norms for vectors and matrices can be found on almost every book on matrix theory or numerical analysis. We recall that the most used norms in a finite dimensional space $\boldsymbol{R}^s$ are:

(1) $\|v\|_1 = \sum_{i=1}^{s} |v_i|,$

(2) $\|v\|_2 = \left(\sum_{i=1}^{s} v_i^2\right)^{1/2},$

(3) $\|v\|_\infty = \max_{1 \le i \le s} |v_i|.$

By means of each vector norm, one defines consistent matrix norm as follows

$$\|A\| = \sup_{x \neq 0} \frac{\|Ax\|}{\|x\|}. \tag{A.5.1}$$

The corresponding matrix norms are

(1′) $\|A\|_1 = \max_{1 \le j \le s} \sum_{i=1}^{s} |a_{ij}|,$

(2′) $\|A\|_2 = (\rho(A^H A))^{1/2},$

(3′) $\|A\|_\infty = \max_{1 \le i \le s} \sum_{j=1}^{s} |a_{ij}|,$

where A is any $s \times s$ complex matrix, A^H is the transpose conjugate of A, and $\rho(A)$ is the spectral radius of A.

Given a nonsingular matrix T, one can define other norms starting from the previous ones, namely, $\|v\|_T = \|Tv\|$ and the related consistent matrix norms $\|A\|_T = \|T^{-1}AT\|$. If T is a unitary matrix and the norm chosen is $\|\cdot\|_2$, it follows that

$$\|A\| = \|A\|_T. \tag{A.5.2}$$

For all the previous defined norms the consistency relations

$$\|Ax\| \leq \|A\| \, \|x\|, \tag{A.5.3}$$

and

$$\|AB\| \leq \|A\| \, \|B\|, \tag{A.5.4}$$

where A, B, x are arbitrary, hold.

Theorem A.5.1. *For every consistent matrix norm one has*

$$\rho(A) \leq \|A\|. \tag{A.5.5}$$

Proof. Let λ be an eigenvalue of A and v the associated eigenvector. Then we have

$$|\lambda| \, \|v\| \leq \|Av\| \leq \|A\| \, \|v\|,$$

from which (A.5.3) follows. ■

Corollary A.5.1. *If* $\|A\| < 1$, *then the series*

$$\sum_{i=0}^{\infty} A^i \tag{A.5.6}$$

converges to $(I - A)^{-1}$, *and* $\|(I - A)^{-1}\| < 1/(1 - \|A\|)$.

Proof. From Theorem A.5.1 it follows that $\rho(A) < 1$ and the results follow from Theorem A.1.6 (see also exercise A.1.1). ■

Another simple consequence is the following (known as Banach Lemma).

Corollary A.5.2. *Let* A, B *be* $s \times s$ *complex matrices such that* $\|A^{-1}\| \leq \alpha$, $\|A - B\| \leq \beta$, *with* $\alpha\beta < 1$. *Then* B^{-1} *exists and* $\|B^{-1}\| < \alpha/(1 - \alpha\beta)$.

Proof. Let be $C = A^{-1}(A - B)$. By hypothesis $\|C\| < \alpha\beta < 1$. Then $(I - C)^{-1}$ exists and $\|(I - C)^{-1}\| \leq 1/(1 - \alpha\beta)$. But $B^{-1} = (I - c)^{-1}A^{-1}$ and then $\|B^{-1}\| \leq \alpha/(1 - \alpha\beta)$. ■

Theorem A.5.2. *Given a matrix* A *and* $\varepsilon > 0$, *there exists a norm* $\|A\|$ *such that* $\|A\| \leq \rho(A) + \varepsilon$.

Proof. Consider the matrix $A' = \frac{1}{\varepsilon} A$ and let X be the matrix that transforms A' into the Jordan form (see section A.4). Then

$$\frac{1}{\varepsilon} X^{-1}AX = J_\varepsilon \equiv \frac{1}{\varepsilon} D + H,$$

where D is a diagonal matrix having on the diagonal the eigenvalues of A and H is defined by (A.2.19)′. For any one of the norms (1′), (2′), (3′). $\|H\| = 1$, and $\|D\| \leq \rho(A)$. It follows that $\|X^{-1}AX\| \equiv \|A\|_X \leq \|D\| + \varepsilon\|H\| \leq \rho(A) + \varepsilon$, which completes the proof. ■

A.6. Nonnegative Matrices

In many applications one has to consider special classes of matrices. We define some of the needed notions.

Definition A.6.1. An $s \times s$ matrix A is said to be

(1) positive ($A > 0$) if $a_{ij} > 0$ for all indices,
(2) nonnegative ($A \geq 0$) if $a_{ij} \geq 0$ for all indices.

A similar definition holds for vectors $x \in \boldsymbol{R}^s$ considered as matrices $s \times 1$.

Definition A.6.2. An $s \times s$ nonnegative matrix A is said to be reducible if there exists a permutation matrix P such that

$$PAP^T = \begin{pmatrix} B & 0 \\ C & D \end{pmatrix} \tag{A.6.1}$$

where B is an $r \times r$ matrix ($r < s$) and D an $(s - r) \times (s - r)$ matrix. Since $P^T = P^{-1}$, it follows that the eigenvalues of B form a subset of the eigenvalues of A.

Definition A.6.3. A nonnegative matrix A, which is not reducible is said to be irreducible.

Of course if $A > 0$ it is irreducible.

Proposition A.6.1. *If A is reducible then all its powers are reducible.*

Proof. It is enough to consider that

$$PA^2P^T = PAP^TPAP = \begin{pmatrix} B^2 & 0 \\ G & D^2 \end{pmatrix}$$

and proceed by induction. ■

Definition A.6.4. An irreducible matrix is said to be primitive if there exists an $m > 0$ such that $A^m > 0$.

Definition A.6.5. An irreducible matrix A is said to be cyclic if it is not primitive.

It is possible to show that the cyclic matrices can be transformed, by means of a permutation matrix P, to the form

$$PAP^T = \begin{pmatrix} 0 & & \cdots & G_r \\ G_1 & 0 & & 0 \\ 0 & G_2 & & \vdots \\ \vdots & & & \\ 0 & \cdots & G_{r-1} & 0 \end{pmatrix}. \tag{A.6.2}$$

In case when A is irreducible we have the following result, due to Perron–Frobenius.

Theorem A.6.2. *If $A > 0$ (or $A \geq 0$ and primitive) then there exist $\lambda_0 \in \mathbf{R}^+$ and $x_0 \geq 0$ such that*

(a) $Ax_0 = \lambda_0 x_0$,
(b) *if $\lambda \neq \lambda_0$ is an eigenvalue of A then $|\lambda| < \lambda_0$,*
(c) *λ_0 is simple.*

If $A \geq 0$ but not primitive, then (a) *is still valid but $|\lambda| \leq \lambda_0$ in* (b).

Theorem A.6.3. *Let $A \geq 0$ be irreducible and λ_0 its spectral radius. Then the matrix $(\lambda I - A)^{-1}$ exists and is positive if $|\lambda| > \lambda_0$.*

Proof. Suppose $|\lambda| > \lambda_0$. The matrix $A^* = \lambda^{-1}A$ has eigenvalues inside the unit circle. Hence it follows (see Corollary A2.1) that $(I - A^*)^{-1}$ is given by

$$(I - A^*)^{-1} = \sum_{i=0}^{\infty} A^{*i}$$

from which it follows

$$(\lambda I - A)^{-1} = \lambda^{-1}(I - A^*)^{-1} = \lambda^{-1} \sum_{i=0}^{\infty} (\lambda^{-1}A)^i. \tag{A.6.3}$$

The second part of the proof is left as an exercise. ∎

Appendix B—The Schur Criterium

We have seen in Chapters 2 and 3 that the asymptotic stability problem for autonomous linear difference equations is reduced to the problem of establishing when a polynomial has all the roots inside the unit disk in the complex plane. This problem can be solved by using the Routh method (see for example [152] and [12]). In this Appendix we shall present the Schur criterium, the one most-used for such a problem.

Let $p(z) = \sum_{i=0}^{k} p_i z^i$ be a polynomial of degree k. The coefficients p_i can be complex numbers. Let $q(z) = \sum_{i=0}^{k} \bar{p}_i z^{k-i}$, ($\bar{p}_i$ are the conjugates of p_i), be the reciprocal complex polynomial: $q(z) = z^k \bar{p}(z^{-1})$. Let S be the set of all the Schur polynomials (see Definition (2.6.4)). Consider the polynomial of degree $k - 1$:

$$p^{(1)}(z) = \frac{\bar{p}_k p(z) - p_0 q(z)}{z}.$$

It is easy to see that $p^{(1)}(z) = \sum_{i=1}^{k} (\bar{p}_k p_i - p_0 \bar{p}_{k-i}) z^{i-1}$.

Theorem B.1. *$p(z) \in S$ iff* (a) $|p_0| < |p_k|$ *and $p^{(1)}(z) \in S$.*

Proof. Suppose $p(z) \in S$ and let $z_1, z_2, \ldots, z_k$ be its roots. Then

$$|p_0| = \left| p_k \prod_{i=1}^{k} z_i \right| < |p_k|$$

and condition (a) is verified. On the unit circle $|z| = 1$ one has

$$|q(z)| = \left| \sum_{i=0}^{k} \bar{p} z^{k-i} \right| = \left| \sum_{i=0}^{k} \bar{p}_i z^{-i} \right| = \left| \sum_{i=0}^{k} \bar{p}_i \bar{z}^i \right| = |\overline{p(z)}|.$$

Since condition (a) is verified, we get

$$|\bar{p}_k p(z)| > |\bar{p}_0 p(z)| = |\bar{p}_0 \overline{p(z)}| = |p_0 q(z)| = |-p_0 q(z)|.$$

Applying Rouche's theorem (see [76]) it follows that the polynomial $\bar{p}_k p(z)$ and $\bar{p}_k q(z) - p_0 q(z) = zp^{(1)}(z)$ have the same number of roots in $|z| < 1$, that means that $p^{(1)}(z) \in S$.

Suppose now that $p^{(1)}(z) \in S$ and $|p_0| < |p_k|$. It follows that on $|z| = 1$, $|\bar{p}_k p(z)| > |-p_0 q(z)|$ and again by Rouche's theorem the polynomial $\bar{p}_k p(z)$ has n roots inside the unit disk, that is $p(z) \in S$.

The previous theorem permits to define an algorithm to check recursively if a polynomial is a Schur polynomial. The algorithm is very easily implemented on a computer.

Next theorem, which is similar to the previous one, gives the possibility of finding the number of roots inside the unit disk. Consider the polynomial of degree $k - 1$

$$Tp(z) \equiv \bar{p}_0 p(z) - p_k q(z) = \sum_{i=0}^{k-1} (\bar{p}_0 p_i - p_k \bar{p}_{k-i}) z^i.$$

The polynomial $Tp(z)$ is called the Schur transform of $p(z)$. The transformation can be iterated by defining $T^s p = T(T^{s-1}p)$, for $s = 2, 3, \ldots, k$. Let $\gamma_s = T^s p(0)$, $s = 1, 2, \ldots, k$. ■

Theorem B.2. *Let $\gamma_s \neq 0$, $s = 1, 2, \ldots, k$, and let $s_1, s_2, \ldots, s_m$ an increasing sequence of indices for which $\gamma_{s_j} < 0$. Then the number of roots inside the unit disk is given by $h(p) = \sum_{j=1}^{m} (-1)^{j-1}(k + 1 - s_j)$.*

Proof. See Henrici [76]. ■

Analogous results can be given for Von-Neumann's polynomials. Let $\boldsymbol{N}$ be such a set.

Theorem B.3. *A polynomial $p(z)$ is a Von-Neumann polynomial if either*

(1) $|p_0| < |p_k|$ *and* (2) $p^{(1)}(z) \in \boldsymbol{N}$

or

(1′) $p^{(1)}(z) \equiv 0$ *and* (2′) $p'(z) \in S$.

Proof. See [113]. ■

Appendix C—Chebyshev Polynomials

C.1. Definitions

The solutions of the second order linear difference equation

$$y_{n+2} - 2zy_{n+1} + y_n = 0, \tag{C.1}$$

where $z \in \boldsymbol{C}$, corresponding to the initial conditions

$$y_0 = 1, \qquad y_1 = z, \tag{C.2}$$

and

$$y_{-1} = 0, \qquad y_0 = 1, \tag{C.3}$$

are polynomials as functions of z and are called Chebyshev polynomials of first and second kind respectively. They are denoted by $T_n(z)$ and $U_n(z)$.

We list the first five of them below.

$$\begin{aligned}
T_0(z) &= 1 & U_{-1} &= 0\\
T_1(z) &= z & U_0(z) &= 1\\
T_2(z) &= 2z^2 - 1 & U_1(z) &= 2z\\
T_3(z) &= 4z^3 - 3z & U_2(z) &= 4z^2 - 1\\
T_4(z) &= 8z^4 - 8z^2 + 1 & U_3(z) &= 8z^3 - 4z.
\end{aligned}$$

Since the Casorati determinant

$$K(0) = \begin{vmatrix} T_0(z) & U_{-1}(z) \\ T_1(z) & U_0(z) \end{vmatrix}$$

is equal to 1, the general solution of (C.1) can be written as

$$y_n(z) = c_1 T_n(z) + c_2 U_{n-1}(z). \tag{C.4}$$

Let w_1 and w_2 be the roots of the characteristic polynomial associate to (C.1), that is,

$$w^2 - 2zw + 1 = 0. \tag{C.5}$$

It is easy to express $T_n(z)$ and $U_n(z)$ in terms of w_1^n and w_2^n. In fact one obtains, considering that $w_2 = w_1^{-1}$,

$$T_n(z) = \frac{w_1^n + w_1^{-n}}{2} = \cosh n \log(z + (z^2 - 1)^{1/2}), \tag{C.6}$$

$$\begin{aligned} U_n(z) &= \frac{1}{(z^2-1)^{1/2}} \frac{w_1^{n+1} - w_1^{-n-1}}{2} \\ &= \frac{1}{(z^2-1)^{1/2}} \sinh(n+1) \log(z + (z^2 - 1)^{1/2}). \end{aligned} \tag{C.7}$$

For $z \in [-1, 1]$, by setting $z = \cos\theta$, it follows that $w_1 = e^{i\theta}$ and

$$T_n(z) = \cosh n \log e^{i\theta} = \cosh ni\theta = \cos n\theta,$$

$$U_n(z) = \frac{\sin(n+1)\theta}{\sin\theta},$$

which are the classical Chebyshev polynomials.

C.2. Properties of $T_n(z)$ and $U_n(z)$

One easily proves that the roots of $T_n(z)$ and $U_{n-1}(z)$ are

$$z_k = \cos\frac{2k+1}{n}\frac{\pi}{2}, \qquad k = 0, 1, \ldots, n-1, \tag{C.8}$$

and

$$z_k = \cos\frac{k\pi}{n}, \qquad k = 1, 2, \ldots, n-1, \tag{C.9}$$

respectively. Other properties of $T_n(z)$ and $U_n(z)$ are the following, which can be verified easily.

(1) $w_1^n = T_n(z) + (T_n^2(z) - 1)^{1/2}$,
(2) $T_n(z) = T_{-n}(z)$, (symmetry)
(3) $T_{jn}(z) = T_j(T_n(z))$, (semigroup property)
(4) $U_{jn-1}(z) = U_{j-1}(T_n(z))$,
(5) $U_{-n} = -U_{n-1}(z)$,
(6) $T_{n-1}(z) - zT_n(z) = (1 - z^2)U_{n-1}(z)$,
(7) $U_{n-1}(z) - zU_n(z) = -T_{n+1}(z)$,
(8) $U_{n+j}(z) + U_{n-j}(z) = 2T_j(z)U_n(z)$.

From the last property one can derive many others. For example,

(9) $U_{n+j-1}(z) + U_{n-j-1}(z) = 2T_j(z)U_{n-1}(z)$,
(10) $U_{n+j}(z) + U_{n-j-2}(z) = 2T_{j+1}(z)U_{n-1}(z)$,
(11) $2T_n(z)U_n(z) = U_{2n}(z) + 1$,
(12) $2T_{n+1}(z)U_{n-1}(z) = U_{2n}(z) - 1$,
(13) $2T_n(z) = U_n(z) - U_{n-2}(z)$.

Among the properties of Chebyshev polynomials the following has a fundamental importance in approximation theory.

(14) Let P_n be the set of all n degree polynomials having as leading coefficient 1. Then for any $p(z) \in P_n$,

$$\max_{-1 \le z \le 1} |2^{1-n}T_n(z)| \le \max_{-1 \le z \le 1} |p(z)|.$$

The proof of this property is not difficult and can be found in several books on numerical analysis.

(15) $T_n(z)$ and $U_n(z)$, as function of z, satisfy the differential equations

$$T_n'(z) = nU_{n-1}(z); \tag{C.10}$$

$$U_n'(z) = -(z^2-1)^{-1}[zU_n(z) - (n+1)T_{n+1}(z)],$$

$$(1-z^2)T_n''(z) - zT_n'(z) + n^2T_n(z) = 0, \tag{C.11}$$

$$(1-z^2)U_n''(z) - 3zU_n'(z) + n(n+2)U_n(z) = 0. \tag{C.12}$$

The first two are consequences of the definitions (C.6) and (C.7) and the other can be derived from them.

Finally $T_n(z)$ and $U_n(z)$ satisfy the orthogonal properties.

(16) $$\frac{2}{\pi}\int_{-1}^{1} \frac{T_n(z)T_m(z)}{1-z^2}\,dz = \begin{cases} 2\delta_{nm} & \text{for } n = 0, \\ \delta_{nm} & \text{for } n > 0, \end{cases}$$

(17) $$\frac{2}{\pi}\int_{-1}^{1} (1-z^2)^{1/2}U_n(z)U_m(z)\,dz = \delta_{nm},$$

(18) $$\sum_{j=1}^{n-1} T_k(z_j)T_m(z_j) + \frac{1}{2}[T_k(-1)T_m(-1) + T_k(1)T_m(1)]$$

$$= \begin{cases} \frac{n}{2}\delta_{km}, & k \ne 0, \quad k \ne m, \\ n\delta_{km}, & k = 0, \quad k = n, \end{cases}$$

where $z_j = \cos\frac{j\pi}{n}$, $j = 1, \ldots, n-1$.

Solutions to the Problems

Chapter 1

1.1. From (1.2.12) $y_5 = E^4 y_1 = \sum_{j=0}^{4} \binom{4}{j} \Delta^j y_1$ and from the scheme

	Δ^0	Δ^1	Δ^2	Δ^3	Δ^4
y_1	-2				
y_2	-2	0			
y_3	0	2	2		
y_4	4	4	2	0	0

we get $y_5 = 10$.

1.3. The first result of problem 1.2 can also be written

$$\Delta\, e^{i(ax+b)} = 2\, e^{i(ax+b+a/2)}\left(\frac{e^{ia/2} - e^{-ia/2}}{2}\right) = 2i \sin\frac{a}{2}\, e^{i(ax+b+a/2)}.$$

By taking the real and imaginary parts one gets the result.

1.4. Using the Sterling transform one has $j^3 = j^{(1)} + 3j^{(2)} + j^{(3)}$ and $\Delta^{-1} j^3 = \frac{j^{(2)}}{2} + j^{(3)} + \frac{j^{(4)}}{4}$ from which one obtains:

$$\sum_{j=0}^{n} j^3 = \Delta^{-1} j^3 \Big|_{j=0}^{j=n+1} = \frac{n^2(n+1)^2}{4}.$$

1.5. From the result of problem 1.3 we have

$$\Delta^{-1}\cos(ax+b+a/2) = \frac{\sin(ax+b)}{2\sin a/2}.$$

Setting $a = q$, $b = -q/2$ and $x = n$, we have:

$$\Delta^{-1}\cos qn = \frac{\sin(qn - q/2)}{2\sin q/2}$$

and

$$\sum_{i=1}^{n}\cos qi = \Delta^{-1}\cos qi\Big|_{i=1}^{i=n+1} = \frac{\sin[(n+1)q - q/2] - \sin q/2}{2\sin q/2}.$$

1.7. From $x^{(n+x-n)} = x^{(n)}(x-n)^{(x-n)}$ and $x^{(x)} = \Gamma(x+1)$, $(x-n)^{(x-n)} = \Gamma(x-n+1)$ we have $x^{(n)} = \dfrac{\Gamma(x+1)}{\Gamma(x+n+1)}$.

1.10. Using the Sterling transformation, $p(x)$ can be written in the form $p(x) = \sum_{i=0}^{k} a_i x^{(i)}$. Applying $\Delta, \Delta^2, \ldots, \Delta^k$ and putting $x = 0$ one gets $a_i = \dfrac{\Delta^i p(0)}{i!}$.

1.11. One has:

$$\begin{aligned}
\sum_{j=0}^{q-n}(-1)^j\binom{q}{n+j} &= \sum_{j=0}^{q-n-1}(-1)^j\binom{q}{n+j} + (-1)^{q-n}\\
&= \sum_{j=0}^{q-n-1}(-1)^j\left[\binom{q-1}{n+j} + \binom{q-1}{n+j-1}\right]\\
&\quad + (-1)^{q-n}\\
&= \sum_{j=0}^{q-n-1}(-1)^j\binom{q-1}{n+j} + \sum_{j=0}^{q-n-1}(-1)^j\binom{q-1}{q-1}\\
&\quad \times (-1)^{q-n}\\
&= \sum_{j=0}^{q-n-1}(-1)^j\binom{q-1}{n+j} + \sum_{j=0}^{q-n}(-1)^j\binom{q-1}{n+j-1}\\
&= \sum_{j=0}^{q-n-1}(-1)^j\binom{q-1}{n+j} + \sum_{j=1}^{q-n}(-1)^j\binom{q-1}{n+j-1}\\
&\quad + \binom{q-1}{n-1}.
\end{aligned}$$

By setting $j-1=i$ in the second term, it is easily seen that the two sums cancel.

1.12. One has:

$$S(q,l,n)=\sum_{j=1}^{q-n-1}(-1)^{j-l}\binom{q}{n+j}\binom{j}{l}+(-1)^{q-n-l}\binom{q-n}{l}$$

$$=\sum_{j=l}^{q-n-1}(-1)^{j-l}\binom{q-1}{n+j}\binom{j}{l}$$

$$+\sum_{j=l}^{q-n-1}(-1)^{j-l}\binom{q-1}{n+j-1}\binom{j}{l}$$

$$+(-1)^{q-n-l}\binom{q-1}{q-1}\binom{q-n}{l}$$

$$=\sum_{j=l}^{q-n-1}(-1)^{j-l}\binom{q-1}{n+j}\binom{j}{l}$$

$$+\sum_{j=l+1}^{q-n}(-1)^{j-l}\binom{q-1}{n+j-1}\binom{j}{l}+\binom{q-1}{n+l-1}$$

$$=\sum_{j=l}^{q-n-1}(-1)^{j-l}\binom{q-1}{n+j}\left[\binom{j}{l}-\binom{j+1}{l}\right]$$

$$+\binom{q-1}{n+l-1}=\binom{q-1}{n+l-1}.$$

1.23. The solutions are $y_n = cn - c^2$.

1.24. Letting $y_n = z_n^{-1}$ the equation becomes $z_{n+1} = z_n + 1$.

1.25. Letting $n = 2^m$, $y_m = \boldsymbol{C}(2^m)$ and $g_{m-1} = f(2^m)$, one has $y_m = 2y_{m-1} + f(2^m) = 2y_{m-1} + g_{m-1}$, $y_1 = 2$. The solution is $y_m = 2^{m-1}y_1 + \sum_{s=1}^{m-1} g_s 2^{m-s-1}$.

1.26. Letting $z_n = A^{1/2}\cot y_n$ one has $y_{n+1} = 2y_n$.

1.27A. From $\log y_{n+1} = 2\log y_n$ one obtains $\log y_n = 2^n \log y_0$ and then $y_n = y_0^{2^n}$.

1.27B. By substituting $y_n = \dfrac{1-z_n}{2}$ the equation becomes $z_{n+1} = z_n^2$, from which one obtains $y_n = \frac{1}{2}[1-(1-2y_0)^{2^n}]$.

1.27C. Let $z_n = 1 - y_n$ and obtain the equation of problem 1.26.

1.28. $y_{n+1}(1+y_n^2) + y_n = k$.

1.30. From the result of problem 1.29 one has

$$u_n \leq \delta \exp\left(H^{1/2}M \sum_{j=0}^{n-1} \frac{1}{(n-j)^{1/2}} \right) + \delta \sum_{s=0}^{n-1} \exp\left(h^{1/2}M \sum_{r=s+1}^{n=1} \frac{1}{(n-r)^{1/2}} \right).$$

It is known that $\sum_{k=1}^{n} k^{-1/2} < 2n^{1/2} - 1$. Then one has

$$u_n \leq \delta\left((\exp(h^{1/2}M(2n^{1/2}-1)) + \sum_{s=0}^{n-1} \exp(h^{1/2}M(2(n-s-1)^{1/2}-1) \right)$$

$$\leq \delta\, e^{-h^{1/2}M}\left(e^{2h^{1/2}Mn^{1/2}} + \sum_{s=0}^{n-1} e^{2h^{1/2}M(n-s-1)^{1/2}} \right).$$

1.31. Let $V_n = c + \sum_{s=n+1}^{\infty} k_s V_s$. One has $y_n \leq V_n$ and $V_n = V_{n+1} + k_{n+1}V_{n+1}$, that is $V_n = \prod_{s=n+1}^{t} (1+k_s)V_t \leq V_t \exp(\sum_{s=n+1}^{t} k_s)$ and for $t \to \infty$, $y_n \leq V_\infty \exp(\sum_{s=n+1}^{\infty} k_s) = c\exp(\sum_{s=n+1}^{\infty} k_s)$.

1.32. For $a < 1$ one obtains $y_{n+1}^2 < y_n^2 + b_n(y_n + y_{n+1})$ and then $y_{n+1} < y_n + b_n$, from which $y_n \leq y_0 + \sum_{j=0}^{n-1} b_j$. For $a > 1$ one has $y_{n+1}^2 - ay_n^2 \leq b_n(y_n + y_{n+1})$ and then $y_{n+1} - a^{1/2}y_n \leq b_n \dfrac{y_n + y_{n+1}}{a^{1/2}y_n + y_{n+1}} < b_n$, from which one obtains $y_n < (a^{1/2})^n y_0 + \sum_{j=0}^{n-1} (a^{1/2})^{n-j-1} b_j$.

Chapter 2

2.8c. Write the equation (2.3.1) in the form $\Delta y_{n+1} = y_n$ and apply Δ^{-1} to both sides.

2.13c. In (2.5.10) one imposes $q(x) = \bar{p}'(x)$ from which it follows $y_0 p_0 = p_1$, $y_0 p_1 + p_0 y_1 = 2p_2, \ldots$.

2.14. Let $\varepsilon_{i+1} = S_{i+1} - \bar{S}_{i+1}$, $r_i = a_{i+1} - \bar{a}_{i+1}$, $\delta_i = 10^{-t}(\bar{s}_i + a_{i+1})$. One has $\varepsilon_{i+1} = \varepsilon_i + r_i - \delta_i$, from which $|\varepsilon_n| \leq k_n + 10^{-t}|\sum_{i=0}^{n-1} \delta_i|$. From this one deduces that one must keep $|\sum \delta_i|$ as small as possible and this can be achieved summing first the smaller a_i in absolute value. If $|\bar{a}_i| < a$, then $|S_i| \leq ia$. Finally one has $|\varepsilon_N| \leq k_N + 10^{-t}a\sum_{i=0}^{N-1} i \leq k_N + 10^{-t}a\dfrac{N(N-1)}{2}$ showing that the error grows like N^2.

2.19 From (2.5.18) we have

$$\sum_{i=0}^{\infty} y_i z^i = \frac{\sum_{n=0}^{\infty} q_n^{(2)} z^{n+2}}{\frac{1}{2}z^2 - \frac{3}{2}z + 1},$$

where $q_0^{(2)} = q_1^{(2)} = 0$ and $q_n = 1/n$. The roots of denominator are 1 and 2. In the unit circle it can be written:

$$\left(\frac{1}{2}z^2 - \frac{3}{2}z + 2\right)^{-1} = \sum_{i=0}^{\infty} \gamma_i z_i$$

with γ_i bounded and $\lim_{i=0} \sum_0^n \gamma_i = \infty$ (why?). Then one has

$$\sum_{i=0}^{\infty} \gamma_i z^i \sum_{n=0}^{\infty} q_n^{(2)} z^{n+2} = \sum_{i=0}^{\infty} c_i z^{i+2},$$

where $c_i = \sum_{n=2}^{i} q_n^{(2)} \gamma_{i-n}$. Then $y_n = c_{n-2} = \sum_{j=0}^{n-2} \dfrac{1}{j+1} \gamma_{n-j-2} \leq \operatorname{Max}_{1 \leq s \leq n} \sum_{j=1}^{s} \gamma_j$. The last inequality follows from the result of problem 1.13a.

2.20. Taking $f = 0$, the error equation becomes $\rho(E)e_n = \rho(1)$ whose solution is the sum of the general solution of the homogeneous equation and a particular solution. The general solution is given by (2.3.14) from which it follows that $\rho(z)$ must be a Von Neumann polynomial.

2.21. Taking $f = 0$ as in the previous problem and $e_0 = e_1 = \ldots = e_{k-1} = 0$, we have (see 2.4.11) $e_n = \rho(1)/\rho(1)$ and this cannot be zero for $n \to \infty$ $(nh \leq T)$ unless $\rho(1) = 0$. Similarly for $f \equiv 1$: $\rho(E)e_n = h(\rho'(1) - \sigma(1))$.

2.24. Use Theorems B.1 and B.2 of Appendix B. For case a) one finds $D = [-i, i]$.

Chapter 3

3.7. Exchanging with care the sum in the expression $\sum_{i=0}^{k} p_i(n) \sum_{j=n_0}^{n-i} H(n+k-i, j) g_i$, one has $\sum_{j=n}^{n-i} g_i \sum_{i=0}^{k} p_i H(n+k-i, j) + g_n = g_n$. (Remember that some values of H are zero).

3.11. $|z_s| \neq 1$, $s = 1, 2$.

3.12. It depends on the roots of $z^2 + p_1 z + p_2$. It will be

$$H(n+1) + p_1 H(n) + p_2 H(n-1) = 0 \qquad \text{for } n \neq 0$$

$$H(1) + p_1 H(0) + p_2 H(-1) = 1.$$

According to the values of the roots, several cases are possible. For example:

(a) $|z_1| < 1, |z_2| > 1$

$$H(n) = \begin{cases} (z_1 + p_1 + p_2 z^{-1})^{-1} z_2^n & \text{for } n \le 0 \\ (z_1 + p_1 + p_2 z_2^{-1})^{-1} z_1^n & \text{for } n \ge 0 \end{cases}$$

(b) $|z_1| < 1$, $|z_2| < 1$, $H(n) = (z_1 - z_2)^{-1}(z_1^n - z_2^n)$ for $n \ge 0$ and $H(n) = 0$ for $n \le 0$.

3.13. $H(n-j) = \dfrac{\sin(n-j-1)t}{\sin t}$; $y_n = \sum_{j=0}^{n-2} \dfrac{\sin(n-j-1)t}{\sin t} g_j$.

3.14. Let y_n^i the nth element of the ith column of the inverse matrix C_N^{-1}. By multiplying C_N and its inverse one shows that the columns of C_N^{-1} are solutions of the boundary value problems:

$$\gamma y_{n-1}^{(i)} + \alpha y_n^{(i)} + \beta y_{n+1}^{(i)} = \delta_n^i, \qquad n = 2, 3, \ldots, N-1$$

$$\alpha y_1^{(i)} + \beta y_2^{(i)} = \delta_1^i$$

$$\gamma y_{N-1}^{(i)} + \alpha y_N^{(i)} = \delta_N^i$$

for $i = 1, 2, \ldots, N$. Let z_1 and z_2 be the roots of $\gamma + \alpha z + \beta z^2 = 0$. The solutions are:

$$y_n^{(i)} = \begin{cases} -\dfrac{\beta^{-1}}{H(N+2)} H(N-i+2)H(n+1) & \text{for } n \le i \\ -\dfrac{\beta^{-1}}{H(N+2)} H(N-n+2)H(i+1)(z_1 z_2)^{n-i} & \text{for } n \ge i, \end{cases}$$

where $H(n) = \dfrac{z_2^{n-1} - z_1^{n-1}}{z_2 - z_1}$. One sees that for $|z_1 z_2| \le 1$ and $|z_2| > 1$, the elements remain bounded for $N \to \infty$.

3.15. Suppose that $A(n)$ is a companion matrix, that is of the form (3.3.3). $A^T(n)$ is not of the same form and then the components of x_n, given by (3.4.9) are not successive values of the solution of a scalar equation. Now consider the matrix

$$V(n) = \begin{pmatrix} -p_k(n) & -p_{k-1}(n) & \cdots & -p_1(n) \\ 0 & -p_k(n+1) & \cdots & -p_2(n+1) \\ \vdots & & & \\ 0 & & \cdots & -p_k(n+k) \end{pmatrix}.$$

One easily verifies that $x_n^T = \tilde{x}_{n+1}^T V(n+1)$, where $\tilde{x}_n^T = (t_n, t_{n+1}, \ldots, t_{n+k-1})$ and t_n is the last component of x_n. For this new vector $\tilde{x}_n$ one has $\tilde{x}_n = x_{n-1}^T V^{-1}(n) = x_n^T A(n) V^{-1}(n) = \tilde{x}_{n+1}^T V(n+1) A(n) V^{-1}(n) \equiv \tilde{x}_{n+1}^T \Omega(n)$. The matrix $\Omega(n) \equiv V(n+1)A(n)V^{-1}(n)$ is given by

$$\Omega(n) = \begin{pmatrix} p_1(n+1) & 1 & 0 & \cdots & 0 \\ p_2(n+2) & 0 & 0 & \cdots & \\ \vdots & & & & 1 \\ p_k(n+k) & 0 & & \cdots & 0 \end{pmatrix},$$

which is the companion form for the adjoint equation defined in Section 2.1.

3.20. From the definition, one has

$$G(j,j) = -\Phi(j,0)Q^{-1} \textstyle\sum_{i=0}^{N} L_i \Phi(n_i, j+1) T(j+1, n_i)$$

and

$$G(j+1,j) = \Phi(j+1, j+1) - A\Phi(j,0)Q^{-1} \times \textstyle\sum_{i=0}^{N} L_i \Phi(n_i, j+1) T(j+1, n_i) = I + AG(j,j).$$

3.21. One has

$$\begin{aligned} \sum_{i=0}^{N} L_i G(n_i, j) &= \sum_{i=0}^{N} L_i \Phi(n_i, j+1) T(j+1, n_i) \\ &\quad - \sum_{i=0}^{N} L_i \Phi(n_i, 0) Q^{-1} \\ &\quad \times \sum_{s=0}^{N} L_s \Phi(n_s, j+1) T(j+1, n_s) \\ &= \sum_{i=0}^{N} L_s \Phi(n_s, j+1) T(j+1, n_i) \\ &\quad - \sum_{s=0}^{N} L_s \Phi(n_s, j+1) T(j+1, n_s) = 0. \end{aligned}$$

3.22. From

$$\begin{aligned} y_{n+1} &= \Phi(n+1, 0)Q^{-1}w + \sum_{s=0}^{n_N-1} G(n+1, s) b_s \\ &= A\Phi(n,0)Q^{-1}w + \sum_{\substack{s=0 \\ s\neq n}}^{n_N-1} AG(n,s)b_s + b_n + AG(n,n)b_n \\ &= A[\Phi(n,0)Q^{-1}w + \sum_{s=0}^{n_N-1} G(n,s)b_s] + b_n = Ay_n + b_n. \end{aligned}$$

For what concerns the boundary conditions one has:

$$\sum_{i=0}^{N} L_i y_{n_i} = \sum_{i=0}^{N} L_i \Phi(n_i, 0) Q^{-1} w + \sum_{i=0}^{N} \sum_{s=0}^{n_N-1} L_i G(n_i, s) b_s$$

$$= w + \sum_{i=0}^{N} \sum_{s=0}^{n_N-1} L_i \Bigg(\Phi(n_i, s+1) T(s+1, n_i)$$

$$- \Phi(n_i, 0) Q^{-1} \sum_{j=0}^{N} L_j \Phi(n_j, s+1) T(s+1, n_j) \Bigg) b_s$$

$$= w + \sum_{s=0}^{n_N-1} \Bigg(\sum_{i=0}^{N} L_i \Phi(n_i, s+1) T(s+1, n_i)$$

$$- \sum_{j=0}^{N} L_j \Phi(n_j, s+1) T(s+1, n_j) \Bigg) b_s = w.$$

Chapter 4

4.3. The system is symmetric with respect to the origin: $g_i(-x, -y) = -g_i(x, y)$ for $i = 1, 2$. This allows us to consider only the upper half plane. According to the signs of g_1 and g_2, let us consider the following sets:

$A = \{(x, y) | y \geq 2x\}$ where $g_1(x, y) \geq 0, g_2(x, y) > 0$

$B = \{(x, y) | y < 2x \text{ and } x^2(y - x) + y^5 \geq 0\}$ where $g_1(x, y) \geq 0, g_2(x, y) < 0$

$C = \{(x, y) | x^2(y - x) + y^5 < 0\}$ where $g_1(x, y) < 0, g_2(x, y) < 0.$

It is clear that for $(x_k, y_k) \in A$, $\Delta x_k \geq 0$, $\Delta y_k \geq 0$, that is both the sequences x_k, y_k are not decreasing. If they remain in A, they will never cross the line $y = 2x$ and the ratio $\dfrac{g_2(x, y)}{g_1(x, y)}$ has to be greater than 2 for all k. This means that $\dfrac{y^2(y - 2x)}{x^2(y - x) + y^5} > 2$, which is impossible because the y in the denominator has degree larger than the y in the numerator. The sequences must cross the line $y = 2x$ and enter in B. In a similar way it can be shown that they cannot remain in

B. They will enter in C, where both Δx_k and Δy_k are negative and the sequences are decreasing. Now if $y_k > 0$ it follows that

$$y_{k+1} = y_k + \frac{y_k^2(y_k - 2x_k)}{r_k^2 + r_k^6} = \frac{y_k(r_k^2 + r_k^6 + y_k^2 - 2x_k y_k)}{r_k^2 + r_k^6}$$

$$= \frac{y_k((x_k - y_k)^2 + y_k^2 + r_k^6)}{r_k^2 + r_k^6} > 0$$

and similarly for x_k. This shows that the sequences must remain in C and must converge to a point where both Δx_k and Δy_k are zero, which is the origin.

Starting from $(-\delta, \delta)$ for small positive δ, it is easy to check that in the following iterations the points (x_k, y_k) have increasing distance from the origin until they reach the regions B or C, showing that the origin is unstable.

4.5. The solution is

$$y(n, n_0, y_0) = \frac{\log(n_0 + 2)}{\log(n + 2)} y_0.$$

The series $\sum_{n=n_0}^{\infty} |y(n, n_0, y_0)|$ does not converge, showing that the origin is not l_1-stable.

4.9. In this case D is the set of all positive numbers. If $y_0 \in D$, then $y_n \in D$ for all n. Consider $V(y) = y/(1 + y^2)$. It is $V(y) \geq 0$ and

$$\Delta V(y_n) = \frac{y_n(1 - y_n)^3(y_n - 1)}{(1 + y_n^2)(1 + y_n^4)} = -W(y_n) < 0.$$

The set E is $\{0, 1\}$ and $W(x) \to 0$ for $x \to \infty$. Then, according to the theorem y_n is either unbounded or tends to E. In fact the solution is $y_n = y_0^{(-2)^n}$ and it tends to zero if $y_0 < 1$ and n even and it is unbounded for n odd. If $y_0 = 1$ then $y_n = 1$ for all n.

4.10. The eigenvalues of A are $\lambda_1 = \frac{1}{2}$ with multiplicity $s - 1$ and $\lambda_2 = (1 - s)/2$. They can be obtained easily by considering that A is a circulant matrix and the eigenvalues are the values assumed by the polynomial $\rho(z) = -\frac{1}{2}(z + z^2 + \ldots + z^{s-1})$ on the sth roots of the unity. There is global asymptotic stability for $s = 2$, only stability for $s = 3$ and instability for $s > 3$.

4.11. In order to have $V_{n+1} \geq 0$, it must be $V_n - \omega(V_n) \geq 0$. If u and v are such that $u \geq n$, $u - \omega(u) \geq 0$, $v - \omega(v) \geq 0$, and $u - v \geq \omega(u) - \omega(v)$ one has $u - \omega(u) \geq v - \omega(v)$. The solution satisfies $u_n = u_0 - \sum_{j=0}^{n-1} \omega(u_j)$. Since ω is increasing and u_n must remain positive, it follows that $u_j \to 0$.

4.12. Let $x \in \Omega(y_0)$, there exists a sequence $n_i \to \infty$ such that $y(n_i, n_0, y_0) \to x$. But $y(n_{i+1}, n_0, y_0) = f(y(n_i, n_0, y_0))$ and $\lim_{n_i \to \infty} y(n_{i+1}, n_0, y_0) = f(x)$, showing that $f(x) \in \Omega(y_0)$ and that $\Omega(y_0)$ is invariant. Now let y_k be a sequence in $\Omega(y_0)$ converging to y. We shall prove that $y \in \Omega(y_0)$. For each index k, there is a sequence $m_i^k \to \infty$ such that $y(m_i^k, n_0, y_0) \to y_k$. Suppose for simplicity that $\text{dist}(y_k, y(m_i^k, n_0, y_0)) < k^{-1}$ and $m_i^k \geq k$ for $i \geq k$. Consider the sequence $m_k = m_k^k$. Then $\text{dist}(y, y(m_k, n_0, y_0)) \leq \text{dist}(y, y_k) + k^{-1}$, which implies $\text{dist}(y, y(m_k, n_0, y_0)) \to 0$ and then $y \in \Omega(y_0)$.

Chapter 5

5.2. From the mean value theorem one has

$$F(x) - F(y) - F'(x)(y - x) = \int_0^1 [F'(x + s(y - x)) - F'(x)]\, ds(y - x);$$

$$\|F(x) - F(y) - F'(x)(y - x)\| \leq \gamma \int_0^1 s\, ds \|y - x\|^2.$$

5.4. Let $\alpha_0 = \frac{1}{2}\beta\gamma$. One has

$$\|x_{n+1} - x_n\| \leq \alpha_0 \|x_n - x_{n-1}\|^2$$

and

$$\|x_{k+m} - x_k\| \leq \sum_{j=k}^{k+m-1} \|x_{j+1} - x_j\| \leq \sum_{j=1}^{m} \alpha_0^{2^j - 1} \|x_k - x_{k-1}\|^{2^j}$$

$$\leq \|x_k - x_{k-1}\|^2 \sum_{j=1}^{m} \alpha_0^{2^j-1} \|x_k - x_{k-1}\|^{2^j - 2}$$

$$\leq \|x_k - x_{k-1}\|^2 \sum_{j=1}^{m} \alpha_0^{2^j-1} u_{k-1}^{2^j-2}$$

$$\leq \|x_k - x_{k-1}\|^2 \sum_{j=1} \alpha_0^{2^j-1} \eta^{2^j-2} (\alpha^{2^k-1})^{2^j-1}$$

$$= \|x_k - x_{k-1}\|^2 \alpha_0 \sum_{j=1}^{\infty} (\alpha_0 \eta)^{2^j-2} (\alpha^{2^k-1})^{2^j-2}$$

$$= \|x_k - x_{k-1}\|^2 \alpha_0 \sum_{j=1}^{\infty} \alpha^{2^j-2} (\alpha^{2^k-1})^{2^j-2}$$

$$= \|x_k - x_{k-1}\|^2 \alpha_0 \sum_{j=1}^{\infty} (\alpha^{2^k})^{2^j-2} = \|x_k - x_{k-1}\|^2 \alpha_0 \frac{1}{1 - \alpha^{2^k}}.$$

5.5. Consider the equation $u_n = \frac{\Delta t_n}{\Delta t_{n-1}} u_{n-1}$. One has $\frac{u_n}{\Delta t_n} = \frac{u_{n-1}}{\Delta t_{n-1}}$, that is $\frac{u_n}{\Delta t_n}$ is constant and must maintain its initial value u_1/t_1. Then $u_n = \Delta t_n(u_1/t_1)$. It follows then that $\Delta x_n \leq u_n$. Moreover $\frac{\|\Delta x_n\|}{\Delta t_n}$ is decreasing, that is, for $m \geq n$, $\frac{\|\Delta x_m\|}{\Delta t_n} \leq \frac{\|\Delta x_n\|}{\Delta t_n}$ and

$$\|x_{n+p} - x_n\| \leq \sum_{j=0}^{p-1} \|x_{n+j+1} - x_{n+j}\|$$

$$\leq \sum_{j=1}^{p-1} \Delta t_{n+j} \frac{\|\Delta x_{n+j}\|}{\Delta t_{n+j}} \leq \sum_{j=1}^{p-1} \Delta t_{n+j} \frac{\|\Delta x_n\|}{\Delta t_n}$$

$$= \frac{t_{n+p} - t_n}{t_{n+1} - t_n} \|\Delta x_n\|$$

from which it follows that

$$\lim_{p\to\infty} \|x_{n+p} - x_n\| \leq \frac{t^* - t_n}{t_{n+1} - t_n} \|\Delta x_n\|.$$

Being $\|\Delta x_n\| \leq u_n$ one has

$$\|x^* - x_n\| \leq \frac{(t^* - t_n)u_1}{t_1}.$$

5.6. In this case the comparison equation is

$$u_n = \frac{\Delta t_n}{(\Delta t_{n-1})^\gamma} u_{n-1}^\gamma,$$

whose solution is $u_n = \Delta t_n \left(\frac{u_1}{t_1}\right)^{\gamma^{n-1}}$. As before one has

$$\|x^* - x_n\| \leq (t^* - t_n)\left(\frac{\|\Delta x_n\|}{\Delta t_n}\right)^\gamma \leq (t^* - t_n)\left(\frac{u_1}{t_1}\right)^{\gamma^n}.$$

5.7. Let $y_n = 1 - z_n$. The equation becomes $y_{n+1} = \frac{1}{2}\left(y_n + \frac{1 - 2z_0}{y_n}\right)$ and apply the result of problem 1.26.

5.8. From the given solution one has $z_0 = 1 - (1 - 2z_0)^{1/2} \coth k$ from which $k = \log \theta^{-1/2}$ and then

$$z_n = 1 - \frac{1 + \theta^{2^n}}{1 - \theta^{2^n}} (1 - 2z_0)^{1/2}$$

from which

$$\|x^* - x_n\| \le \frac{1}{\beta\gamma}(z^* - z_n) = \frac{2}{\beta\gamma}(1 - 2z_0)^{1/2}\frac{\theta^{2^n}}{1 - \theta^{2^n}}.$$

5.12. The equation (5.5.2) is the homogeneous equation related to (5.5.1) when one takes

$$x_n = \frac{y_{n-1}}{y_n}.$$

Imposing that (5.5.4) satisfies the nonhomogeneous equation one gets (5.5.3). In fact one has $y_{n+1} - 2\delta y_n + x_n y_n + z_n = g_n$ from which

$$y_n = \frac{y_{n+1}}{2\delta - x_n} + \frac{z_n - g_n}{2\delta - x_n} \equiv x_{n+1}y_{n+1} + z_{n+1}.$$

Chapter 6

6.1. Consider the term $\sum_{j=0}^{n-1} A^{n-j-1} W_j$ and the decomposition (A2.2). It follows that $A = Z_{11} + \sum_{k=2}^{d} (\lambda_k Z_{k1} + Z_{k2}) \equiv S + S_1$, where d is the number of distinct eigenvalues of A. Using the properties of the component matrices Z_{kj} it is seen that $S^j = S$ for all j and $\lim_{j\to\infty} S_1^j = 0$. Moreover $A^j = S^j + S_1^j$. The sum $\sum_{j=0}^{n-1} A^{n-j-1} W_j$ becomes $S\sum_{j=0}^{n-1} W_j + \sum_{j=0}^{n-1} S_1^{n-j-1} W_j$. The quantity SW_j is called essential local error. If the errors are such that $SW_j = 0$ then one can proceed as usual.

6.5. Applying the theorem B3 one has $p^{(1)}(z) = -4\,\mathrm{Re}q$ and $p'(z) = 2(z - q)$. It follows that $p(z) \in \boldsymbol{N}$ iff $\mathrm{Re}q = 0$ and $|q| < 1$.

6.8. Rewrite the equation (6.1.15) such that the linear autonomous matrix A is the companion matrix of $\sigma(z)$, obtaining $E_{n+1} = (A + B_n)E_n + W_n$, where $B_n = \Phi_k b^T$ with b is now the vector with components.

$$b_j = \frac{\alpha_j - \beta_j + h\beta_j(\beta_k c_{n+k} - c_{n+j})}{2 - h\beta_k c_{n+k}}.$$

Find a bound for b_j independent on n and consider that by hypothesis $\|A\| < 1$.

6.9. Consider the nonhomogeneous equation $C_N x^N - \lambda x^N = \delta^{(N)}$, where $\|x^N\| = 1$ for all N and $\|\delta^{(N)}\| = \varepsilon^2$. Componentwise one considers the equation $\beta x_{n-1}^N + (\alpha - \lambda)x_n^N + \gamma x_{n+1}^N = \delta_n^N$ whose solutions are chosen such that the initial condition $(\alpha - \lambda)x_1^N + \gamma x_2^N = 0$ is satisfied. In the hypothesis made the solution will diverge and (6.7.1) will not be satisfied.

6.12. The matrix A_N is the matrix representing the centered second order discretization, that is

$$A_N = \begin{pmatrix} -2 & 1 & 0 & & \cdots & 0 \\ 1 & -2 & 1 & 0 & \cdots & 0 \\ 0 & & & & & 0 \\ \vdots & & \ddots & \ddots & \ddots & \vdots \\ & & & & & 1 \\ 0 & \cdots & 0 & & 1 & -2 \end{pmatrix}_{N \times N}$$

$$D_{2N}^2 = \begin{pmatrix} \Delta x^2 A_N & 0 \\ 0 & \Delta x^2 A_N \end{pmatrix},$$

which is symmetric. The spectrum of $(\Delta x^2)^{-1} D_{2N}^2$ is $S = \{-2 + 2\cos\theta,\ 0 \le \theta < \pi\}$. The spectrum of D_{2N} is then $S^{1/2}$ (see proposition 6.7.3), which is an interval of the imaginary axis. The midpoint rule can be used because its region of absolute stability lies on the imaginary axis and Δt can be chosen appropriately in order that the region contain $S^{1/2}$.

Chapter 7

7.2. From 2.5.10 one has $X(z) = \dfrac{q(z)}{1 - \sum_{i=1}^{M} a_i z^i}$. The problem reduces to the study of the roots of the denominator. They are the reciprocal of the characteristic roots and they are outside the circle $B(0, \lambda_1^{-1})$. The series $X(z) = \sum_{n=0}^{\infty} x_1(n) z^n$ is then convergent in this circle. To see if $z_1 = \lambda_1^{-1}$ is less or equal to 1, one considers that $1 - \sum_{i=1}^{M} a_i z^i$ is monotone decreasing on the real line from 1 to $-\infty$ and the z_1 will be less or greater than 1 according to the sign of $1 - \sum a_i$. If this quantity is positive, then $z_1 > 1$, otherwise $z_1 < 1$. In the first case one has

$$\sum_{n=0}^{\infty} x_1(n) = \frac{1}{1 - \sum a_i}$$

and $x_1(n) \to 0$. In the second case $x_1(n) \to \infty$. If $1 - \sum a_i = 0$ then by Theorem 2.5.2 the solution $x_1(n)$ is unbounded.

7.6. By definition $f[x, f(x)] = \dfrac{f(x) - f^{(2)}(x)}{x - y}$. If $f^{(2)}(x) = x$ then the fraction assumes the value -1, and vice versa.

7.9. The model is

$$P_1 = \rho_0 P_0$$

$$(n+1)P_{n+1} - (\rho_n + n)P_n + \rho_{n-1}P_{n-1} = 0$$

$$NP_N - \rho_{N-1}P_{N-1} = 0.$$

To obtain the solution, observe that

$$(n+1)P_{n+1} - \rho_n P_n = nP_n - \rho_{n-1}P_{n-1}.$$

That is $nP_n - \rho_{n-1}P_{n-1} = K$, where K is a constant. From the first equation one sees that $K = 0$. One obtains

$$P_n = \frac{\prod_{i=0}^{n-1} \rho_i}{n!} P_0$$

and then from $\sum_{i=1}^{N} P_i = 1$ one obtains P_0.

References

[1] Agarwall, R. P. and Thandapani, E., *On some new discrete inequalities*, Appl. Math. and Computation, **7** (1980), pp. 205, 224.
[2] Agarwall, R. P. and Thandapani, E., *Some inequalities of Gronwall type*, Analele Stjietifice Univ. Iasi., XXVII (1981), pp. 139, 144.
[3] Agarwall, R. P., *On multipoint boundary value problems for discrete equations*, Jour. of Math. Analysis and Appl., **96** (1983), pp. 520-534.
[4] Agarwall, R. P., *Initial value method for discrete boundary value problems*, Jour. of Math. Analysis and Appl., **100** (1984), pp. 513-529.
[5] Agarwall, R. P., *Difference calculus with applications to difference equations*, In W. Walter, General Inequalities **4**.
[6] Albrecht, P., *Numerical treatment of ODE: the theory of A-methods, Num. Math.*, **47** (1985), pp. 59-87.
[7] Alekseev, V. M., *An estimate for the perturbation of the solution of ordinary differential equation*, (Russian) Vestu. Mosk. Univ. Ser I, Math. Med., **2** (1961), pp. 28-36.
[8] Arneodo, A., Ferrero, P. and Tresser, C., *Sharkovskii order for appearance of superstable cycles: An elementary proof*, Comm. Pure and Appl. Math. XXXVII (1984), pp. 13-17.
[9] Atkinson, F. V., *Discrete and continuous boundary value problems*, Academic Press (1964).
[10] Bakhvalov, N. S., *Numerical Methods*, MIR, Moscow, (1977).
[11] Barna, B., *Uber die diverenzpunbe des Newtonches verbfahrens zur bestinemmung von wurzelu algebraischen*, Publications Mathematical Debreceu, **4** (1956), pp. 384-397.
[12] Barnett, S. and Silijak, D. D., *Routh algorithm, a centennial survey*, SIAM Review, **19** (1977), pp. 472-489.
[13] Bermand, A. and Plemmons, R., *Nonnegative matrices in the Mathematical Sciences*, Academic Press, New York, (1979).
[14] Bernussou, J., *Point Mapping Stability*, Pergamon, (1979).
[15] Beyn, W. J., *Discrete Green's functions and strong stability properties of the finite difference method*, Appl. Analysis, (1982), pp. 73-98.
[16] Brand, L., *Differential and Difference Equations*, J. Wiley, (1966).
[17] Burrage, K. and Butcher, J. C., *Stability criteria of implicit Runge-Kutta methods*, SIAM JNA, **15** (1979), pp. 46-57.

[18] Burrage, K. and Butcher, J.C., *Nonlinear stability for a general class of differential equation methods*, BIT, **20** (1980), pp. 185–203.

[19] Bus, J. C. P., *Numerical solution of systems of nonlinear equations*, Mathematical Centre Tracts, Amsterdam, (1980).

[20] Butcher, J.C., *A stability property for implicit Runge-Kutta methods*, BIT, **15** (1975), pp. 358–361.

[21] Cash, J. R., *Finite Reccurrence Relations*, Academic Press, (1976).

[22] Cash, J. R., *An extension of Olver's method for the numerical solution of linear recurrence relations*, Math of Comp., **32** (1978), pp. 497–510.

[23] Cash, J. R., *Stable Recursions*, Academic Press, London, (1979).

[24] Cash, J. R., *A note on the solution of linear recurrence relations*, Num. Math., **34** (1980), pp. 371–386.

[25] Clenshaw, C. W., *A note on the summation of Chebyshev series*, MTAC, **9** (1955), pp. 118–120.

[26] Coffman, C. V., *Asymptotic behavior of solutions of ordinary differential equations*, Trans. Am. Math. Soc., **110** (1964). pp. 22–51.

[27] Collatz, L., *Einige anwendungen funktionanalytischer methoden in der praktischen analysis*, Z. Angen. Math. Phys., **41** (1953) pp. 327–357.

[28] Collatz, L., *Functional Analysis and Numerical Mathematics*, Academic Press, (1966).

[29] Collet, P. and Eckmann J. P., *Iterated Maps on the Interval and Dynamical Systems*, Birkhauser, London, (1980).

[30] Corduneanu, C., *Principles of Differential and Integral Equations*, Chelsea, New York, (1977).

[31] Corduneanu, C., *Almost periodic discrete processes*, Libertas Math., **2** (1982), pp. 159–169.

[32] Dahlquist, G. and Bjork, A., *Numerical Methods*, Prentice-Hall, (1974).

[33] Dahlquist, G., *A special stability problem for linear multistep methods*, BIT, **3** (1963), pp. 27–43.

[34] Dahlquist, G., *Error analysis for a class of methods for stiff nonlinear initial value problems*, Num. Anal., Dundee, Springer Lect. Notes in Math., **506** (1975), pp. 60–74.

[35] Dahlquist, G., *On stability and error analysis for stiff nonlinear problems*, Part I, Report Trita-NA-7508, (1975).

[36] Dahlquist, G., *G-stability is equivalent to A-stability*, BIT, **18** (1978), pp. 384–401.

[37] Dahlquist, G., *On the local and lobal errors of one-leg methods*, Report, TRITA-NA-8110, (1981).

[38] Dahlquist, G., *Some comments on stability and error analysis for stiff nonlinear differential systems*, preprint, NADA Stockholm, (1983).

[39] Dahlquist, G., Liniger W., and Nevanlinna O., *Stability of two step methods for variable integration steps*, SIAM JNA, **20** (1983), pp. 1071–1085.

[40] Deuflhard, P. and Heindl G., *Affine invariant convergence theorems for Newton's method and extensions to related methods*, SIAM JNA, **16** (1979), pp. 1–10.

[41] Deuflhard, P., *On algorithm for the summation of certain special functions*, Computing, **17** (1976), pp. 37–48.

[42] Deuflhard, P., *A summation technique for minimal solutions of linear homogeneous difference equations*, Computing, **18** (1977), pp. 1–13.

[43] Diamond, P., *Finite stability domains for difference equations*, Jour. Austral. Soc., **22A** (1976), pp. 177–181.

[44] Diamond, P., *Discrete Liapunov function with $V > 0$*, Jour. Austral. Soc., **20B** (1978), pp. 280–284.

[45] Di Lena, G. and Trigiante D., *On the stability and convergence of lines method*, Rend. di Mat., **3** (1982), pp. 113–126.

[46] Driver, R. D., *Note on a paper of Halanay on stability of finite difference equations*, Arch. Rat. Mech., **18** (1965), pp. 241–243.

[47] Fielder, M. and Ptak, V., *On matrices with nonpositive off-diagonal elements and positive principal minors*, Czechoslovakia Math. Jour., **12** (1962), pp. 382–400.

[48] Fielder, M. and Ptak, V., *Some generalizations of positive definiteness and monotonicity*, Num. Math., **9** (1966), pp. 163–172.

[49] Fort, T., *Finite Differences and Difference Equations in the Real Domain*, Oxford, (1948).

[50] Gandolfo, G., *Mathematical Methods and Models in Economics Dynamics*, North-Holland, Amsterdam, (1971).

[51] Gantmacher, F. R., *The Theory of Matrices*, Vol. 1-2, Chelsea, (1959).

[52] Gautschi, W., *Computational aspects of three terms recurrence relations*, SIAM Rev., **9** (1967), pp. 24–82.

[53] Gautschi, W., *Minimal solutions of three term reccurrence relations and orthogonal polynomials*, Math. of Computation, **36** (1981), pp. 547–554.

[54] Gear, G. W., *Numerical Initial Value Problems in Ordinary Differential Equations*, Prentice-Hall, (1971).

[55] Gear, G. W. and Tu K. W., *The effect of variable mesh size on the stability of multistep methods*, SIAM JNA, **1** (1974), pp. 1025–1043.

[56] Gekerel, E., *Discretization methods for stable initial value problems*, Lecture Notes in Math., Springer, (1984).

[57] Gelfond, A. O., *Calcul des Difference Finies*, Dunod, Paris, (1963).

[58] Gelfond, A. P., *Calculus of Finite Differences*, Hindusten Publishing Corp., Delhi, (1971).

[59] Godunov, S. K. and Ryabenki, V. S., *Theory of Difference Schemes*, North-Holland, (1984).

[60] Goldberg, S., *Introduction to Difference Equations*, J. Wiley, New York, (1958).

[61] Gordon, S. P., *Stability and summability of solutions of difference equations*, Math. Syst. Theory, **5** (1971), pp. 56–75.

[62] Grobner, W., *Die Lie-Reiben und Ihre Anwendungen*, Berlin, DVW, (1960).

[63] Grujic, L. J. T. and Siljak, D. D., *Exponential stability of large scale discrete systems*, J. Control, **19** (1976), pp. 481–491.

[64] Guckenheimer, J., Oster, G., and Ipaktchi, A., *The dynamics of density dependent population models*, J. Math. Biology, **4** (1977), pp. 101–147.

[65] Hahn, W., *Stability of Motion*, Springer, Berlin, (1967).

[66] Haight, F. A., *Mathematidal Theories of Traffic Flow*, Academic Press, (1963).

[67] Hairer, E. and Wanner, G., *Algebraically stable and implementable Runge-Kutta methods of higher order*, SIAM JNA, **18** (1981), pp. 1098–1108.

[68] Halanay, A. and Wexler, D., *Teoria Calitative a Sistemlor cu Impulsuri*, Bucharest, (1968).

[69] Halanay, A., *Solution periodiques et presque-periodiques des systems d'equationes aux difference finies*, Arch. Rat. Mech., **12** (1963), pp. 134–149.

[70] Halanay, A., *Quelques questions de la théorie de la stabilité pour les systèmes aux différences finies*, Arch. Rat. Mech., **12** (1963), pp. 150–154.

[71] Hang-Chin, Lau and Pou-Yah, Wu, *Error bounds for Newton type process on Banach spaces*, Num. Math., **39** (1982), pp. 175–193.

[72] Hartman, P. and Wintner, A., *On the spectre of Toeplitz matrices*, Am. J. of Math., **72** (1950), pp. 359–366.

[73] Hartman, P., *Difference equations: disconjuancy, principal solutions, Green's functions, complete monotonicity*, Trans. Am. Math. Soc., **246** (1978), pp. 1–30.

[74] Hebrici, P., *Discrete Variable Methods for Ordinary Differential Equations*, J. Wiley, (1962).

[75] Henrici, P., *Error Propagation for Difference Methods*, J. Wiley, (1963).

[76] Henrici, P., *Applied and Computational Complex Analysis*, Vol. 1, J. Wiley, New York, (1974).
[77] Hildebrand, F. B., *Finite Difference Equations and Simulations*, Prentice-Hall, (1968).
[78] Hildebrand, F. B., *Methods of Applied Mathematics*, Prentice-Hall, (1952).
[79] Hoppensteadt, F. C. and Hyman, J. M., *Periodic solutions of a logistic difference equation*, SIAM JAM **32** (1977), pp. 73–81.
[80] Hoppensteadt, F. C., *Mathematical Methods of Population Biology*, Courant Inst. of Math. Science, (1976).
[81] Hurt, J., *Some stability theorems for ordinary difference equations*, SIAM JNA, **4** (1967) pp. 582–596.
[82] Jones, G. S., *Fundamental Inequalities for discrete and discontinuous functional equations*, J. Soc. Ind. Appl. Math., **12** (1964), pp. 43–57.
[83] Jordan, C., *Calculus of Fnite Difference*, Chelsea, New York, (1950).
[84] Kalman, E. and Bertram, J. E., *Control System analysis and design via the 'second method' of Lyapunov, Part II, Discrete time Syst.*, Trans. ASME, Ser. D.J. Basic Enr., **82** (1960), pp. 394–400.
[85] Kannan, R. and Ray, M. B., *Monotone iterative methods for nonlinear equations involving a non-invertible linear part*, Num. Math., **45** (1984), pp. 219–225.
[86] Kato, T., *Perturbation Theory for Linear Operators*, Springer, (1966).
[87] Khavanin, M. and Lakshmikantham, V., *The method of mixed monotony and first order differential systems*, Nonlinear Analysis, **10** (1986), 873–877.
[88] LaSalle, J. P., *The stability of dynamical systems*, Regional Conference Series in Applied Mathematics, SIAM (1979).
[89] Ladde, G. S., Lakshmikantham, V., and Vatsala, A. S., *Monotone Iterative Techniques for Nonlinear Differential Equations*, Pitman Publishers Co., (1985).
[90] Lakshmikantham, V. and Leela, S., *Differential and Integral Inequalities*, Academic Press, New York, Vol. I & II, (1969).
[91] Lakshmikantham, V. and Vatsala, A. S., *Method of mixed monotony for nonlinear equations with a singular linear part*, Appl. Math. and Computations, **23** (1987), 235–241.
[92] Lambert, J. A., *Computational Methods in Ordinary Differential Equations*, J. Wiley, (1973).
[93] Lancaster, P., *Theory of Matrices*, Academic Press, (1969).
[94] Levy, H. and Lessmann, F., *Finite Difference Equations*, McMillan, New York, (1961).
[95] Li, T. Y. and Yorke, J.A., *Period three implies chaos*, Am. Math. Monthly, **82** (1975), pp. 985–992.
[96] Liniger, W., *Numerical solution of ordinary and partial differential equations*, Notes of a cource taught at the Suisse Federal Institute of Technology, Lausanne, (1971–1973).
[97] Lorenz, E. N., *The problem of deducing the climate from the governing equations*, TELLUS, **16** (1964), pp. 1–11.
[98] Luenberger, D. G., *Introduction to Dinamic Systems*, J. Wiley, (1979).
[99] Luke, Y. L., *The Special Functions and Their Approximations*, Academic Press, Vol. I, (1969).
[100] Maslovskaya, L. V., *The stability of difference equations*, Diff. Equations, **2** (1966), pp. 608–611.
[101] Maté, A. and Nevai, P., *Sublinear perturbations of the differential equation $y = 0$ and the analogous difference equation*, J. Diff. Eq., **53** (1984), pp. 234–257.
[102] Mattheij, R. M. and Vandersluis, A., *Error estimates for Miller's algorithm*, Num. Math., **26** (1976), pp. 61–78.
[103] Mattheij, R. M., *Characterizations of dominant and dominated solutions of linear recursions*, Num. Math., **35** (1980), pp. 421–442.

[104] Mattheij, R. M., *Stability of block LU-decompositions of the matrices arising from BVP*, SIAM J. Alg. Dis. Math., **5** (1984) pp. 314–331.

[105] Mattheij, R. M., *Accurate estimates for the fundamental solutions of discrete boundary value problems*, J. Math. Anal. and Appl., **101** (1984), pp. 444–464

[106] May, R. M., *Simple mathematical models with very complicated dynamics*, Nature, **261** (1976), pp. 459–467.

[107] May, R. M., *Biological populations with nonoverlapping generations. Stable points, Stable Cucles and Chaos*, Science, **186** (1974), pp. 546–647.

[108] McKee, S., *Gronwall Inequalities*, Zamm., **62** (1982), pp. 429–431.

[109] Meil, G. J., *Majorizing sequences and error bounds for iterative methods*, Math. of Computations, **34** (1960), pp. 185–202.

[110] Mickley, H. S., Sherwood, T. K., and Reed, C. E., *Applied Mathemaitcs in Chemical Engineering*, McGraw-Hill, N. Y., (1967).

[111] Miel, G., *An updated version of the Kantarovich theorem for Newton's method*, Computing, **27** (1981), pp. 237–244.

[112] Miller, J.C. P., *Bessel Functions, Part II*, Math. Tables, Vol X, British Association for the Advacement of Sciences, Cambridge Univ. Press., (1952)

[113] Miller, J. J. H., *On the location of zeros of certain classes of polynomials with applications to numerical analysis*, J. Inst. Math. Applic., **8** (1971), pp. 397–406.

[114] Miller, K. S., *An Introduction to the Calculus of Finite Differences and Difference Equations*, Hold and Company, New York, (1960).

[115] Miller, K. S., *Linear Difference Equations*, Benjamin, New York, (1968).

[116] Milne-Thomson, L. M., *The Calculus of Finite Differences*, McMillan & Co., London, (1933).

[117] Nevanlinna, O. and Odeh, F., *Multiplier techniques for linear multistep methods*, Num. Funct. Anal. and Optimiz., **3** (4) (1981), pp. 377–423.

[118] Nevanlinna, O. and Liniger, W., *Contractive methods for stiff differential equations*, BIT, **19** (1979), pp. 53–72.

[119] Nevanlinna O., *On the behaviour of the lobal errors at infinity in the numerical interaction of stable IVP*, Num. Math., **28** (1977), pp. 445–454.

[120] O'Shea, R, P., *The extention of Zubov method to sampled data control systems described by difference equations*, IEEE, Trans Auto. Conf., **9** (1964), pp. 62–69.

[121] Odeh, F. and Liniger, W., *Non Linear fixed h stability of linear multistep formulas*, J. Math. Anal. Appl., **61** (1977), pp. 691–712.

[122] Olver, F. W. and Sookne, D. J., *Note on backward recurrence algorithms, Math. of Computation*, **26** (1972), pp. 941–947.

[123] Olver, F. W., *Numerical solutions of second order linear difference equations*, Jour. of Research, NBS, **71B** (1967), pp. 111–129.

[124] Olver, F. W., *Bounds for the solution of second-order linear difference equations*, Journal of Research, NBS, **71B** (1967), pp. 161–166.

[125] Ortega, J. M., *Stability of difference equations and convergence of iterative processes*, SIAM JNA, **10** (1973), pp. 268–282

[126] Ortega, J.M. and Rheinboldt, W. C., *Monotone iterations for nonlinear equations with application to Gauss-Seidel methods*, SIAM JNA, **4** (1967), pp. 171–190.

[127] Ortega, J. M. and Rheinboldt, W. C., *Iterative Solution of Nonlinear Equations in Several Variables*, Academic Press, (1970).

[128] Ortega, J. M. and Voigt, G., *Solution of partial differential equations on vector and parallel computers*, SIAM Rev., **27** (1985), pp. 149–240.

[129] Ortega, J. M., *The Newton-Kantarovich theorem*, Am. Math. Monthly, **75** (1968), pp. 658–660.

[130] Ortega, J. M., *Numerical Analysis*, Academic Press, New York, (1972).

[131] Ostrowski, A., *Les points d'atraction et de repulsion pour l'iteration dans l'esace a n dimension*, C. R. Acad. Sciences, Paris, **244** (1957), pp. 288–289.
[132] Ostrowski, A., *Solution of Equations and Systems of Equations*, Academic Press, New York, (1960).
[133] Pachpatte, B. B., *Finite difference inequalities and an extension of Lyabunov method*, Michigan Math. J., **18** (1971), pp. 385–391.
[134] Pachpatte, B. G., *On some fundamental inequalities and its applications in the theory of difference equations*, Ganita, **27** (1976), pp. 1–11.
[135] Pachpatte, B. G., *On some discrete inequalities of Bellman–Bihari type*, Indian J. Pure and Applied Math., **6** (1975), pp. 1479–1487.
[136] Pachpatte, B. G., *On the discrete generalization of Gronwall's inequality*, J. Indian Math. Soc., **37** (1973), pp. 147–156.
[137] Pachpatte, B. G., *Finite difference inequalities and their applications*, Proc. Nat. Acad. Sci., India, **43** (1973), pp. 348–356.
[138] Pasquini, L. and Trigiante, D., *A globally convergent method for simultaneously finding polynomial roots*, Math. of Computation, **44** (1985), 135–150.
[139] Patula, W., *Growth, oscillations and comparison theorems for second order difference equations*, SIAM J. Math. An., **10** (1979) pp. 1272–1279.
[140] Piazza, G. and Trigiante, D., *Propagazione degli errori nella integrazione numerica di equazioni differenziali ordinarie*, Pubbl. IAC III, Roma **120** (1977).
[141] Pollard, J. H., *Mathematical Methods for the Growth of Human Populations*, Cambridge Univ. Press, (1973)
[142] Popenda, J, *Finite difference inequalities*, Fasciculi Math., **13** (1981), pp. 79–87.
[143] Popenda, J., *On the boundness of the solutions of difference equations*, Fasciculi Math., **14** (1985), pp. 101–108.
[144] Potra, F. A., *Sharp error bounds for a class of Newton-like methods*, Libertas Matematica, **5** (1985), pp. 71–84.
[145] Potts, R. B., *Nonlinear difference equations*, Nonlinear Anal. TMA, **6** (1982), pp. 659–665.
[146] Redheffer, R. and Walter, W., *A comparison theorem for difference inequalities*, Jour. of Diff. Eq., **44** (1982), pp. 111–117.
[147] Rheinboldt, W. C., *Unified convergence theory for a class of iterative processes*, SIAM JNA, **5** (1968), pp. 42–63.
[148] Saaty, T. L., *Modern Nonlinear Equations*, Dover, New York, (1981).
[149] Saaty, T. L., *Elements of Queuing Theory with Applications*, Dover, New York, (1961).
[150] Samarski, A. and Niloaiev, E., *Methodes de Resolution des Equation des Mailles*, MIR, Moscow, (1981).
[151] Sand, J., *On one leg and linear multistep formulas with variable stepsize*, Report, TRITA-NA-8112, (1981).
[152] Schelin, C. W., *Counting zeros of real polynomials within the unit disk*, SIAM JNA, **5** (1983), pp. 1023–1031.
[153] Schoenberg, I. J., *Monosplines and Quadrature Formulae*, In T.N.E. Greville, Theory and Applications of Spines Functions, Academic Press, (1969).
[154] Scraton, R. E., *A modification of Miller's recurrence algorithm*, BIT, **12** (1972), pp. 242–251.
[155] Sharkovskii, A. N., *Coexistence of cycles of continuous map of line into itself*, Ukrainian Math. J., **16** (1964), pp. 61–71.
[156] Shintani, H., *Note on Miller's recurrence algorithm, Jour. of Science of the Hiroshima University*, Ser. A. I, **29** (1965), pp. 121–133.
[157] Skeel, R., *Analysis of fixed step-size methods*, SIAM JNA, **13** (1976), pp. 664–685.
[158] Smale, S., *The fundamental theorem of algebra and complexity theory*, Bull. Am. Math. Soc., **4** (1981), pp. 1–36.

[159] Smith, R. A., *Sufficient conditions for stability of a class of difference equations.* Duke Math. Jour., **33** (1966), pp. 725–734.

[160] Spiegl, M. R., *Finite Difference and Difference Equations*, Schaum Series, New York, (1971).

[161] Stephan, P., *A theorem of Sharkovskii on the existence of periodic orbits of continuous endomorphisms of the real line*, Comm. Math. Phys., **54** (1977), pp. 237–248.

[162] Straffin, P. D., *Periodic points of continuous functions*, Math. Mag., **51** (1978), pp. 99–105.

[163] Sugiyama, S., Difference inequalities and their applications to stability problems, Lec-tures Notes in Math., Springer, **243** (1971), pp. 1–15.

[164] Svirezhev, Y. M. and Logofet, D. O., *Stability of Biological Community*, MIR, Moscow, (1983).

[165] Toeplitz, O., *Zur Theorie der Quadratischen Fozunlu von unendlichrillen Verenderlichen. Gïtinger Nachtrichten* (1910), pp. 351–376.

[166] Traub, J. F., *Iterative Methods for the Solution of Equations*,, Prentice-Hall, (1964).

[167] Trigiante, D. and Sivasundaram, S., *A new algorithm for unstable three term recurrence relations*, Appl. Math. and Comp. **22** (1987), pp. 277–289.

[168] Urabe, M., *Nonlinear Autonomous Oscillations, Academic Press*, (1976).

[169] Van der Cruyssen, P., *A reformulation of Olver's algorithm for the numerical solution of second order difference equations*, Num. Math., **32**(1979), pp. 159–166.

[170] Varga, R., *Matrix Iterative Analysis*, Prentice-Hall, (1962).

[171] Wanner, G. and Reitberger, H., *On the perturbation formulas of Grobner and Alekseev*, Bul. Inst. Pol. Iasi, **XIX** (1973), pp. 15–25.

[172] Weinitschke H., *Uber eine kass von iterations verfahren*, Num. Math, **6** (1964), pp. 395–404.

[173] Weissberger, A. (Editor), *Technique of Organic Chemistry, Vol IV. Distillation*, Inter-science, N.Y., (1951).

[174] Willet, D. and Wong, J. S. W., *On the discrete analogues of some generalizations of Gronwall's inequality*, Monatsh. Math., **69** (1965), pp. 362–367.

[175] Wimp, J., *Computation with Recurrence Relations*, Pitman, (1984).

[176] Yamaguti, M. and Ushiki, S., *Discretization and chaos*, C.R. Acad. Sc. Paris, **290** (1980), pp. 637–640.

[177] Yamaguti, M. and Hushiki, S., *Chaos in numerical analysis of ordinary differential equations*, Phisica, **3D** (1981), pp. 618–626.

[178] Yamaguti, M. and Matano, H., *Euler's finite difference scheme and chaos*, Proc. Japan Acad., **55A** (1979), pp. 78–80.

[179] Yamamoto, T., *Error bounds for Newton's iterated, derived from the Kantarovich theorem*, Num. Math., **48** (1986), pp. 91–98.

[180] Zahar, R. V. M., *Mathematical analysis of Miller's algorithm*, Num. Math., **27** (1977), pp. 427–447.

Subject Index

D

E

F

G

H

I

J

K

Mathematics in Science and Engineering

Edited by William F. Ames, *Georgia Institute of Technology*

Recent titles

R. P. Gilbert and J. Buchanan, *First Order Elliptic Systems*
J. W. Jerome, *Approximation of Nonlinear Evolution Systems*
Anthony V. Fiacco, *Introduction to Sensitivity and Stability Analysis in Nonlinar Programming*
Hans Blomberg and Raimo Ylinen, *Algebraic Theory for Multivariable Linear Systems*
T. A. Burton, *Volterra Integral and Differential Equations*
C. J. Harris and J. M. E. Valenca, *The Stability of Input–Output Dynamical Systems*
George Adomian, *Stochastic Systems*
John O'Reilly, *Observers for Linear Systems*
Ram P. Kanwal, *Generalized Functions: Theory and Technique*
Marc Mangel, *Decision and Control in Uncertain Resource Systems*
T. L. Teo and Z. S. Wu, *Computational Methods for Optimizing Distributed Systems*
Yoshimasa Matsuno, *Bilinear Transportation Method*
John L. Casti, *Nonlinear System Theory*
Yoshikazu Sawaragi, Hirotaka Nakayama, and Tetsuzo Tanino, *Theory of Multiobjective Optimization*
Edward J. Haug, Kyung K. Choi, and Vadim Komkov, *Design Sensitivity Analysis of Structural Systems*
T. A. Burton, *Stability and Periodic Solutions of Ordinary and Functional Differential Equations*
V. B. Kolmanovskii and V. R. Nosov, *Stability of Functional Differential Equations*
V. Lakshmikantham and D. Trigiante, *Theory of Difference Equations: Numerical Methods and Applications*